GB & USA
The mothers of invention

Their contribution to the modern world

This edition first published
in October 2019 by

Keith Chamberlain

Copyright © Keith Chamberlain

All rights reserved. No part of this publication may be reproduced in any material form (including scanning, photocopying or storing in any medium by electronic means and whether or not transiently or incidentally to some other use of this publication) without the copyright owner's written permission, except in accordance with the provisions of the Copyright, Designs and Patents Act 1988 or under the terms of a license issued by the copyright licensing agency, 90 Tottenham Court Road, London, W1P 9HE. Applications for the copyright owner's permission to reproduce any part of this publication should be addressed to the publisher.

The moral right of the author has been asserted in accordance with the Copyright, Designs and Patents Act 1988.

Printed and bound in the UK

ISBN: 978-1-9161414-3-8

A CIP catalogue record for this book is available from the British library.

This book can be ordered direct from the publishers at: sales@greentechpublishing.com

www.greentechpublishing.com

Acknowledgements

I would like to thank all the people who have helped and mentored me over the years in my professional life. To those who gave me the passion and desire to explore and research the countries of Great Britain and the United States of America and chart their recent history of innovation and invention, I am forever grateful.

Special thanks must also go to my American associate Dr David Liebers MD, whose command and application of the English language never ceases to amaze me. Finally, to Greentech Publishing and their team who have provided the perfect balance of support, process and guidance in order to produce this first edition.

To Keiko

CONTENTS

Introduction	09
Patent	12
Great Britain and the USA's top 20 inventions and discoveries	15
Agriculture	16
Business Innovation, retail and brands	19
Clock making	28
Communication - Audio - Visual	30
Criminology and Crime	40
Cryptology	43
Digital and Computing world	45
Discovery and establishment of the USA	57
British Commonwealth countries - past and more recent	59
Emergency Services and Devices	66
Engineering	69
Fabric manufacturing and process	88
Food and drink	94
Household and general innovations	101
The Industrial Revolution	119
Industry – Process and innovation	122
Mathematics	132
Military	136
Music and Musical instruments	145
Photography and Cinematics	152
Publishing and authors	156

Science	169
Astronomy	176
Biology	181
Chemistry	184
Medical innovation	191
Medicine	202
Physics	214
Sport	237
Transport	250
Aviation	250
Rail	261
Road	268
Sea	283
Conclusion	288
About the author	292
Chronological index – by year	294

Great Britain & USA
The mothers of invention
Their contribution to the modern world

Introduction

A study by MITI – Japan's equivalent to the British and American Departments of Commerce and Industry — concluded that 54% of the world's most significant inventions of the last 200 years are British. Of the remainder, 25% are American and 9% Japanese.

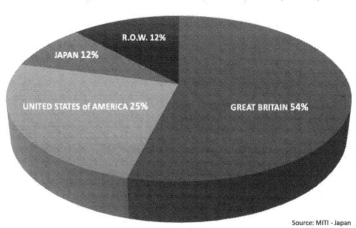

That figure becomes even more astonishing when set against the relative size of Great Britain, the 80th largest country (by area) globally, and the comparative age of the USA, being less than 250 years old.

Yet both countries contribution to the modern world cannot be fully captured by these statistics alone. Though impressive, it leaves out the global reach of both the British and American system in human terms.

The Anglo-American school and university structure produces a culture of intellectual freedom and leadership, requisite for creative output on this scale.

Invention or Innovation?

Anglo-American innovation and invention have been in abundance for centuries. But the word 'innovation' has only come into vogue in recent decades in order to describe the broader concept of how businesses or researchers adopt radically new methods to complement and advance traditional practices. This linguistic sleight of hand has happened in stealth, at the expense of the word 'invention', which strikes the modern ear as primeval or coming from a past century, but not our own. The term 'invention' has quickly disappeared from the commercial lexicon.

The absence of invention in the modern world is easy to miss. On balance, most people see innovation as merely a modern rendering of the word invention. This is founded on the incorrect assumption that they have the same meaning. According to the Oxford Dictionary, invention is: 'the action of creating something (tangible) that has never been made before, or the process of creating something that has never been made before'. Whereas innovation is described in this way: 'to use a new idea or method'. What is true is that they both involve the act of creation.

Nevertheless, there is such a thing as pure invention that stands apart from pure innovation. Pure invention can take the following form: a new product is designed, produced, commercialized, and then sold to people, thus improving their lives. Invention often drives innovation when new inventions help drive changes in existing production practices.

This book is not intended to be the definitive book of Anglo-American inventions. It is not intended to be an encyclopedic or all-encompassing index of every Anglo-American inventive achievement. Rather, it is an entryway. It is a point of reference that will be added to, refined and advanced over the coming years. It is principally about the people of both countries: the imagineers, creators, discoverers and inventors of the small but powerful country of Great Britain and the youthful and hugely ingenious young nation of the USA.

The mothers of Invention

Why write a book on inventions of just two countries, Great Britain and USA? Certainly, there are many other worthy nations that can boast great scientists, inventors and thinkers. Though, what is exceptional is the contrast between Great Britain's diminutive size and the USA's relative age and their standing now, one - Great Britain as the worlds leading 'Soft power' and the USA as the world's dominating 'hard' power. Both countries are irrefutably exceptional and possibly two of the greatest modern nations that the world has ever known. For both countries, their most recent achievements in comparison to its peers, are immense.

The aim of this book is to document a representative sample of these achievements in plain English, so as not to deter our younger readers with technical jargon or granular detail. My hope is that the reader will develop an understanding of the impact and scale of Great Britain and the USA's achievements. Ultimately, I have endeavored to write a chronicle that can both serve as a reference tool for students and become a fixture in every library, and on every coffee table.

During the compilation and writing of this book, I have drawn on a great number of sources, but I accept that there may be errors that are wholly of my own making. I will therefore be pleased to receive any corrections, great or small, along with any significant additions that merit inclusion in later publications. For this I would be eternally grateful. Please send to edits@greentechpublishing.com.

This book is a snapshot. Given the almost interminable inventory of British and US inventions, I made a choice to include the most memorable, distinguished and significant ones in recent history. If I have succeeded, then readers will be inspired to seek out more information and detail from further, more eminent sources. My thesis is simple: Great Britain and the USA are undeniably 'the mothers of invention' in the modern world.

Patent

The most important legal safeguard for an inventor

A short history of the patent

The word patent originates from the Latin, patere, meaning 'to lay open' (or to make accessible for public inspection). It is an abbreviated form of the term letters patent, which was an open document or mechanism supplied by a government or monarch that grants exclusive rights to a person, preluding the modern patent system by several hundreds of years. Comparable permits included land patents, that were generally land rights, issued by early state governments in the USA, including printing rights that were a forerunner of modern copyright.

The importance of a patent

Essentially, if it's worth copying, it's worth safeguarding.

If your invention, design or creation is good enough that someone may copy it, then it is worth investing in IP protection so that you can prevent future copying for a fixed period dependant on the country issuing the patent. A third party may also want to capitalise on your invention by taking it one step further in commercialising your invention. Hindsight is easy, but in practice originators of ideas don't know which of their creations will be successful or become flops and in the short time it takes somebody to adapt a similar concept, it may be too late to apply for a patent to put legal protection in place.

British patent history   1449

Henry VI of England granted the earliest known English patent for an invention created by Flemish-born John of Utynam in 1449, through an open letter marked with the King's Great Seal known as a letter patent.

In Britain, the English patent system then developed from its early medieval roots into the world's first modern patent system that recognised intellectual property in order to encourage invention; this was the fundamental legal basis upon which the Industrial Revolution would develop and flourish.

Through the 16th century, the English crown would regularly exploit the granting of patents for monopolies. Following public uproar, King James 1 of England (VI of Scotland) was compelled to withdraw all remaining patents granted to monopolies and announce that they were only to be used in the future, for "projects of new invention". This was embedded in law through 'the Statute of monopolies' (1624) wherein Parliament limited the power of the crown unambiguously so that the King or Queen could only release patents to the introducers or inventors of original inventions for a fixed term of years. This decree became the basis for future patent law development, not just in England, but across the world.

First U.S. patent issued — 1790

Signed by President George Washington, Samuel Hopkins was issued with the USA's first patent on July 31st, 1790. The patent was awarded for a process that produced potash (pot and pearl ashes), a key ingredient in the making of fertilizer. Since this momentous day, more than six million patents have been issued by the United States Patent and Trademark Office.

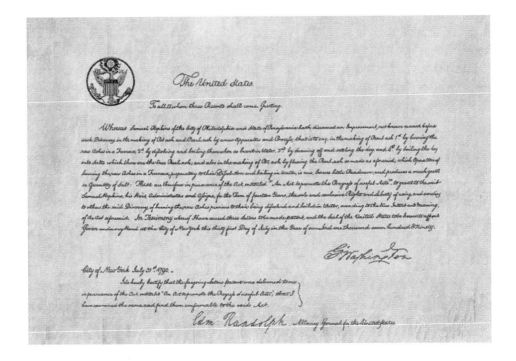

U.S. Congress passed the first patent act on April 10, 1790, entitled "An Act to promote the progress of useful Arts".

In 1836 a major revision to the U.S. patent law was passed. The 1836 law established a far more rigorous application process, including the introduction of an examination system. By the time of the American Civil War, such was the inventiveness of the American people and the success of the US patent system, that about 80,000 patents had been granted.

GB and USA. Top 20 inventions and discoveries

This top 20 of British and American inventions and discoveries is the definitive list from more than six hundred achievements documented in this book. Of course, it is with certainty that many will argue passionately as to what should be considered worthy of placing in the top twenty from such an illustrious collection of truly British and American inventions and innovation. However, I have spent three years researching, compiling and publishing my first phase of both the GB and the USA's disproportionate but truly colossal achievements versus the contribution from the rest of the world. I am therefore using author's prerogative to decide on a ground-breaking top twenty list of achievements that have truly contributed to positive change and progress in the modern world, to set the scene ahead.

 1. English language
 2. The Internet
 3. World Wide Web
 4. Hypodermic syringe
 5. Vaccination
 6. Electric motor and Generator
 7. Space exploration
 8. Television
 9. Mobile phones
 10. Programmable computer
 11. Passenger railway
 12. Chemotherapy
 13. Powered and controlled flight
 14. Jet engine
 15. GPS
 16. Stainless steel
 17. Combine harvester
 18. Electric guitar
 19. Bicycle
 20. Digital camera

Agriculture

 Combine harvester   **1834**

The first combine harvester was invented by Wisconsin based Hiram Moore and was drawn by horses or mules. The harvester was capable of reaping, threshing and sorting cereal grain.

 Baler - round   **1903**

A farm machinery device employed to cut and compress a raked and cut crop such as straw or hay into compact round bales, the baler was truly an agricultural game changer. Co-invented by Ummo F Luebben and his brother Melchior of Sutton, Nebraska, the round hay baler truly transformed the monotonous task of single man baling. The machine essentially gathered the hay at the front of the machine, rolled it into a round bale, then ejected it into the field.

Agriculture

 Gasoline engine – with forward and reverse 1892

The forerunner, donor engine and inspiration of the famous John Deere tractor, was the invention of the world's first successful gasoline-powered combustion engine that could be propelled backwards and forwards. John Froelich created a single cylinder engine that he integrated into the running gear on the chassis of a traction engine. His invention became known as the 'Froelich tractor' and inspired a long line of gasoline engine powered tractors, including the famous two-cylinder John Deere tractor.

 Genetic modification of a plant cell 1982

A team of scientists from the Monsanto Company became the first in the world to introduce a newly discovered gene into the petunia plant. This breakthrough discovery was made public the following year.

 Grain elevator 1842

The first grain elevator was conceived and built by Joseph Dart. His device used a proprietary steam engine for propulsion and later became an invaluable machine in global agriculture, further developed by others later using electricity to replace steam power. Dart who lived most of his life in Buffalo as a lumber trader, was also a founder member of the Buffalo Seminary and pioneer of the buffalo water works.

 Hay baling - automatic 1936

Up to the mid-19th century, cutting hay was a manual task generally using scythes and sickles. Although a stationary baler was invented in the 1850's, it was not until the 1930's that an automatic baler was developed by a company based in Iowa. Ed Nolt, one of their customers took on the challenge of introducing an automatic twine knotter in 1936. New Holland Co. acquired Nolts design under license, and it is believed that he kept his patent. Nolts invention effectively became the world's first commercially successful self-tying hay baler with automatic field collection.

Mechanical reaping machine   **1828**

Patrick Bell was a church minister born 1799, who studied divinities at the University of St Andrews. While working on his father's farm, Bell developed the reaping machine in an effort to speed up the hugely laborious manual task of harvesting. In 1828 his reaping machine, that was pushed and powered by horses, was used on other farms in his district to wide acclaim. Unfortunately for Bell, he never applied for a patent because it was his belief that his creation should benefit everybody. As a consequence, Bell did not make a financial gain on his invention as it enjoyed huge success across the world. In 1831, a patent was issued in the USA to William Manning for a reaper of very similar design.

 Seed drill   **1701**

Born in London, inventor Jethro Tull's seed drill device gouged a straight rut into the ground at a pre-set depth, dropping seeds at regularly spaced intervals. It not only sped up the process of cultivation and crop planting, but also reduced wastage, helping to increase both output and life expectancy of the general population for the first time in history.

 Threshing machine   **1776**

Andrew Meikle was a mechanical engineer who is recognized for inventing a contraption designed to remove the outer husks from grains of wheat, known as 'the Threshing Machine'. This was regarded as one of the foremost inventions in the world of agriculture and its concept is still used in the mechanized world of farming to this day.

 Tractor – powered by internal combustion engine   **1892**

A tractor is designed to provide a high torque and traction at low speed; ideal for towing or pulling agricultural machinery and trailers. Although steam powered tractors had been developed previously, John Froelich's invention made history by being the first tractor, powered by a gasoline powered engine.

Business Innovation, retail and brands

 Amazon 1994

Founded by Jeff Bezos on July 4th, 1994 in Bellevue Washington, Amazon originally started as an online book retailer and marketplace. Realizing that its success could be expanded to other consumer goods such as CDs, clothing, electronics, video games food, toiletries and jewelry. Bezos had unwittingly created what was to become the foremost online shopping company in the world. Bezos cites that his founding principle was to become 'customer obsessed', believing that if it didn't listen to its customers, then it would fail. He capitalized on the unprecedented technological revolution that was to come. Amazon now has more than 270,000 employees globally and generates revenue of more than $240 Billion per year and growing.

Apple 1977

Apple made history on August 2nd, 2018 by becoming the first $1 trillion company in the world. But the story of how it got to this momentous milestone is just as remarkable. Friends, Steve Jobs and Steve Wozniak co-founded Apple in 1977, with the sole purpose of becoming a revolutionary computer hardware company. Their first computers were the Apple 1 and Apple 2, with moderate success and a growing band of loyal customers who loved the functionality and quirky looks of their computers. The pair took the company public in 1980 with Wozniak, the shy mastermind and Jobs the intense visionary. Jobs was ousted by the board in 1985 because of his radical and optimistic vision that was at odds with the rest of the executive.

Whilst away from the cut and thrust of Apple, Jobs founded NeXT high-end computers, that prompted the board in luring him back to head up Apple in the late 1990's. Jobs then set about restructuring the company, taking on board British designer Jony Ive (now Sir Jonathon Ive) and between them, introduced the iPod, iPhone, iPad and iMac. A move that was to transform Apple into a world leader of communication and technology.

 ARM 1990

ARM holdings of Cambridge, GB, is probably the largest tech company that you have never heard of. ARM is an intellectual property inventor, innovator and provider, responsible for creating the architecture used in low-power, high efficiency chips that power most of the world's smartphones, hard drives, and other devices. Often described as the R&D department for the global semiconductor industry. ARM designs chips, licenses their designs, and then accumulates a royalty every time an ARM designed chip is made.

The Apple A4 chip uses an ARM design, then Apple garnishes the ARM base chip with their own custom flare. Used in all Apple phones and almost certainly the forthcoming Apple TV. ARM's designs are in 1.4 billion chips manufactured each quarter, and ARM architecture is in more than 90% of the world's smartphones.

Business innovation, retail and brands

 Assembly line - moving   **1901**

Not to be confused with simply mass production that was invented in Britain during the industrial revolution in the late 16th century, the moving assembly line is a manufacturing process where transposable components are added to a product in a set sequence in order to create a finished product far more quickly than previous traditional methods of manufacturing. The net result of this pioneering process is lower product selling price, lower labor costs and much better parts control. The principal of the moving assembly line was invented by Ransom Olds who used the system to construct the first mass produced automobile, the Oldsmobile Curved Dash, although his groundbreaking achievements are often overlooked by the publicity surrounding his competitor, Henry Ford.

 Barcode   **1952**

Generally, an optical machine-readable image of data. A barcode represents data within the widths of lines and the spacings and thickness of the parallel lines. More recently barcodes have also evolved into patterns of squares, dots and other geometric patterns. The inventor of the barcode was Norman Joseph Woodland.

 Cash register   **1879**

A cash register is a machine used for recording and calculating sales and often for securely storing cash takings within the same unit. The original design revealed the amount tendered on a large front facing dial. As each sale completed, the operator kept a record of transactions through a punched hole paper tape. Its inventor was James Ritty, from Dayton, Ohio. He was granted a patent in the same year of prototype testing, in 1879.

Chain stores - world's first 1848

WH Smith & Son built the world's first chain stores. WH Smith began selling books and other goods at train stations belonging to the London & North-Western Railway, expanding to high street stores in 1905. There are now more than 1500 WH Smith stores in the company's portfolio.

 Credit card 1946

Part of a system of payment processing, a credit card is a small plastic card issued to members of a financial institution that grants a line of credit to the card user, who can borrow money when used as payment or as a cash advance. The credit cards inventor was Flatbush National Bank employee, John C Biggins.

 Mail order 1834

Sir Pryce Pryce-Jones was an entrepreneur who created the world's first mail order business in 1834. Pryce-Jones capitalized on the local wool market and rather than take up farming like his peers, decided to start a business selling woolen flannels locally. When his business grew, he started to advertise in national newspapers and dispatched them by post, first in Great Britain and then overseas. This was the world's first mail order business.

 Mass production C.1700

Mass production was first employed during the start of the industrial revolution in Great Britain in the Derwent Valley. Parts started to be manufactured elsewhere and then assembled in a factory on a facility known as a production or assembly line. This mass production resulted in a standardized product of mid to low quality. By its very nature, the work was very repetitive for the workforce. However, the assembly lines could be stopped as demand dictated and workers for the first time could be laid off, rather than stockpiling goods when demand is low, thus costs were greatly reduced in the process.

Business innovation, retail and brands

 ## McDonalds 1955

Although the very first single store was founded in 1955 by the McDonald brothers in California, it was the dogged determination and vison of Ray Croc, a milkshake machine salesman, who formed the catalyst for expansion and success of the now ubiquitous brand as it stands today. Croc was 51 before he even contemplated entering the restaurant business. Eventually, Ray convinced the McDonald brothers to expand with himself heading up the business expansion division, by adopting a franchise model, based on the first restaurants perfected operations. After a few successful openings, Croc became frustrated by the McDonalds refusal to change and improve the operation and eventually bought the brothers first restaurant in San Bernardino, with its systems and rights to the first few franchises, for the princely sum of $2.7 million cash.

Ray Croc perfected the self-service fast food principle. There was no indoor seating, the menu was limited to a few drinks, fries and hamburgers, all of which were produced in an assembly line operation. He prided himself in being able to process a customer's order in less than a minute. This was to be the foundation for McDonalds success and other fast-food models for the years ahead. Now, McDonalds is the world's largest fast-food burger restaurant company, with more than 37,900 locations in more than 100 countries. What Croc invented was not just a brand or a business, but a symbol of all that America stands for. The moral of his success story was: 'Work hard and you'll make your own success'.

 ## Microsoft 1975

Now, a global computer technology organisation, Microsoft's origins date back to 4th April 1975. Created by Harvard college drop-out Bill Gates and his friend Paul Allen, Microsoft became the world's largest software company and one of the global top 5 companies by value. The company's big break came in 1980 when Gates and Allen formed a partnership with IBM, providing a key operating system for IBM's PC's. Resulting in a royalty payment to Microsoft for every IBM computer sold globally. Then in 1990, when Windows 3.0 was launched, the company sold more than 60 million copies worldwide, effectively making Microsoft the sole supplier of the global PC software standard. The rest is history.

Mini skirt   1966

British fashion designer Mary Quant is generally acknowledged as the person who introduced the mini skirt and coined the phrase by producing short waist skirts that were set about six inches above the knee. Quant never took credit for the invention of the mini skirt but was instead recognised as the person who coined the phrase 'mini' for her short skirt designs.

Rolex - world's largest luxury watch brand   1905

Rolex is recognised as a Swiss high-end watchmaker and ranks as 64th in the world global power brands. However, Rolex started its life in London as a precision watch making collaboration between Alfred Davis and Hans Wilsdorf. Rolex later moved its base to Geneva, Switzerland and is now the world's largest luxury watch brand.

Supermarket   1916

Invented and developed by entrepreneur Clarence Sanders, the supermarket is a self-service store that offers a wide range of produce, departmentalized and paid for at a manned pay station. Although this part of the process today can use unmanned registering and payment. Supermarkets are generally larger than a convenience store and are less personal, with fewer staff per customer. Saunders first supermarket opened in 1916 in Memphis, Tennessee.

Universal Product Code - UPC   1974

Invented by IBM employee, George Laurer, the universal product code or UPC consists of a barcode whose reader scans a 12-digit number along the bar code to enable accurate tracking of goods in and sales out. The UPC also holds shelf price information and is invaluable in either manual cashier checkout or self-pay stations in supermarkets.

 Walmart 1950

The world's largest bricks and mortar retailer by some margin, Walmart operates in 27 countries and owns almost 11,500 stores. Almost sixty percent of its revenue is generated in the USA home market, with its next biggest revenue contributor being the British Asda supermarket brand. Walmart's success is down to one of business history's ultimate operational and logistical successes. The whole organization is dedicated to driving costs out of the supply chain to provide the very best value for money and choice to the customer. This business model has been the company's foundation since it was formed by businessman Sam Walton, when he opened his first store, then called Walton's, in Bentonville, Arkansas in 1950.

Wealth creation 1985-

Britain is the capital of the world for the super-rich, with London boasting more billionaires per head than anywhere else on Earth. Criticized by some, many of these wealth creators are at the heart of the booming economy and as their success and wealth grows, so too does the growth of job and wealth creation for a country that occupies just 1% of the world's total landmass.

"Ninety-nine percent of the failures come from people who have the habit of making excuses"

George Washington Carver

Top 5 GB retail brands (by value 2019)

1. Marks & Spencer
2. Boots Plc
3. Tesco
4. Asda
5. Next

Top 5 GB banking brands (by value 2018)

1. HSBC
2. Lloyds Banking Group
3. Royal Bank of Scotland
4. Barclays Bank
5. Standard Chartered Bank

Top 5 GB Commercial brands (by value 2018)

1. Shell
2. BP
3. Vodafone
4. EE
5. Jaguar Landrover

Source: Bloomberg

Top 5 US retail brands (by value 2019)

1. Walmart
2. The Kroger Co
3. Amazon
4. Costco
5. The Home Depot

Top 5 US banking brands (by value 2018)

1. JP Morgan-Chase
2. Bank of America
3. Wells Fargo
4. Citi
5. US Bancorp

Top 5 US digital, media and internet (by value 2018)

1. Apple
2. Amazon
3. Google
4. Facebook
5. Netflix

Source: Kantar

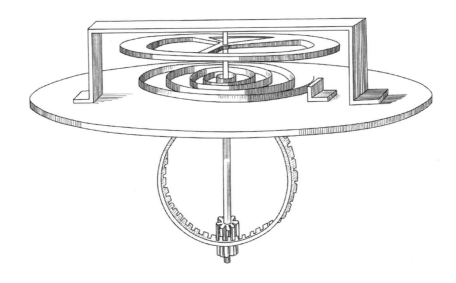

Clock making

 Balance spring   **1660**

Robert Hooke was a British architect and philosopher, educated at Oxford University and born in the Isle of Wight in 1635. Hooke was responsible for many scientific and engineering inventions and theories, but it was his creation of the Balance Spring that revolutionized portable and pocket watches. His invention came about more than 15 years before Christiaan Huygens published his own work on an identical device in 1675.

 Clock Oldest - continually working clock   **1386**

Salisbury Cathedral is recognised as having the oldest working clock in existence. It is a large framed clock, without the common dial, located in the aisle of the great Cathedral. The clock belongs to a group of astronomical clocks from the 14th Century and uses the original mechanism for its operation.

Clock making

 ### Clock - first electric   1841

Alexander Bain was a British engineer and inventor who is credited with designing and making the world's first electric clock. His clock used a pendulum that maintained movement without a clockwork mechanism, by using electromagnetic pulses to maintain momentum. Bain was also famous for installing telegraph lines between Edinburgh and Glasgow.

 ### Clock - first atomic clock   1955

Louis Essen was a British scientist based at the National Physical Laboratory in Teddington, London. He developed the first atomic clock that kept better time than a pendulum or quartz mechanism. Essen's clock was also considerably more constant than the Earth's rotation.

 ### Clock - quartz   1927

Invented and developed by Bell Telephone Laboratory employees, J W Horton and Warren Marrison, the Quartz clock kept extremely accurate time through the use of a quartz crystal oscillating regulator. This breakthrough invention provided mechanical clocks with far better accuracy than sprung wound variants or standard electric clocks that relied on a fluctuating mains frequency.

 ### Wristwatch - first automatic   1923

John Harwood is regarded as the pioneer of the automatic wristwatch. Harwood was a British soldier and a respected watchmaker. After the Great War he founded the Harwood Watch Company. Harwood's aim was to devise a mechanism much better than existing clockwork watches. He noted that the presence of humidity and dust caused severe issues in accuracy of timekeeping in existing designs and so he set out to create a watch that would wind up from the inside of the watch, making it easier to seal. It was the sight of children playing on a seesaw that sparked Harwood's imagination and it set him on the path to inventing the world's first self-winding mechanism: the automatic watch.

Communication – Audio-visual

 Acoustic suspension speaker   **1954**

An acoustic suspension speaker is a type of loudspeaker that moderates distortion of bass frequencies, often experienced with traditional loudspeakers. The acoustic suspension loudspeaker was invented by Edgar Villchur and commercialized by Villchur and Henry Kloss who founded Acoustic Research in Cambridge, Massachusetts.

 Cable television   **1948**

Providing television to customers employing fixed cables or more recently fiber optics to transmit radio frequency signals, rather than receiving TV programming in the traditional method, over-the-air. The co-inventors of cable television (CATV) are Margaret Walson and John Walson from Pennsylvania.

Communication – Audio-Visual

 ### Cordless telephone   1965

A wireless telephone handset, the cordless telephone transmits and communicates using radio waves to and from a base station that is connected to a fixed telephone line. The cordless telephones inventor was Teri Pall but due to lack of marketing and funding, never patented her invention.

 ### Email   1971

A method of creating, transmitting and storing text-based communication between humans, electronic mail, or email as it is better known, was invented and implemented by Ray Tomlinson while working for the US Department of Defense's ARPANET project. Tomlinson is credited as having sent the world's first email on a network and is also recognized for creating the '@' sign.

 ### Emergency calls - 999   1937

Britain's 999 system is the world's oldest emergency call service and was introduced in London on 30th June 1937. Any phone call on the 999 number triggered a special light alarm system ensuring that the incoming 999 calls always took priority at the telephone operation centre. This ground-breaking emergency phone system was quickly adopted throughout the world, some using different numbering systems and is now very much part of the make-up of British life.

 ### Fibre optics   1954

John Tyndall was a British physicist who established that light signals could be sent through a curved stream of water, proving that light could be bowed. Tyndall later demonstrated the same effect by shining light into a pipe filled with water with bends in its length and as the water came out of the pipe into a tank, the water and light curved round in an arc. His discovery paved the way for glass optics that could be bent with minimal deterioration to light travel and output.

Frequency modulation – FM 1933

In radio and telecommunications, FM transports data on a carrier wave by adjusting its frequency. FM's inventor, Edwin Armstrong, from Columbia's Philosophy Hall, decided to create sound by varying the frequency of the radio wave rather than vary the amplitude, creating a distortion free, static free sound compared to the more popular AM at the time. Armstrong was subsequently granted a patent on December 26th, 1933.

 Global positioning system - GPS 1972

GPS is a global navigation system that is based in space, that provides continual three-dimensional, accurate positioning and navigation, regardless of the weather. It can be accessed on land, sea or air, anywhere on or close to Earth via 24 geo-stationary satellites. The co-invention of the global positioning system is credited to a team of engineers who were working for the US Navy, including: Colonel Bradford Parkinson, Bob Rennard, Mel Birnbaum and Jim Spiker.

 Gramophone record 1887

Commonly known as a vinyl record or simply a record, it is a permanent medium capable of analog sound storage on a flat disc with a controlled spiral groove rotating towards the centre point of the record. In response to Edison's cylindrical phonograph, inventor Emile Berliner devised a new standard for recording and playback of sound in the shape of a horizontal disc, known initially as the 'platter'.

 LCD projector 1984

Displaying video, images and computer data, the LCD projector is a modern variant of a light bulb lit overhead projector. The LCD (Liquid Crystal Diode) projector was invented by Gene Dolgoff in 1984 and projects light powered images from a metal-hydride lamp through a prism of dichroic filters, separating light into three polysilicon panels, red, green and blue. Now used the world over from domestic to commercial projector installations.

Communication – Audio-Visual

 **Microphone** 1878

David E Hughes was a British born scientist who is credited with inventing the world's first carbon microphone, essential to the development of the telephone. Hughes' invention was the basis for all future microphones still used today.

 **Mobile phone** 1973

A long-range electronic device that is used for data and voice communication across a cellular network of specialized base stations also known as cell sites. Cellular mobile phones evolved to combat the previously limited FM operated radio telephones that were extremely limited in their performance in many areas. To resolve this, areas were divided into cells so that when users moved between cellular areas, the call would be switched seamlessly between cells. The first mobile call made whilst travelling in a car was made in St Louis, Missouri on June 17th, 1946 but the system was impractical due to its lack of portability, weighing in excess of eighty pounds. The cellular mobile phone was co-invented and perfected by W Rae Young and Douglas H Ring at Bell Labs in 1947. Although the first commercially viable phone call was made to Joel S Engel in April 1973.

 **Morse code** 1836

Invented and developed by Samuel Morse, his invention of coded bleeps (either long or short) used another relatively young medium, the telegraph, to transport the coding from one point to another. The coding represented numbers, letters, punctuation and special characters in any given message. Transmitted by an operator or sender, then transcribed by a decoder or receiver at the other end.

Phonograph   **1877**

Invented by renowned inventor and entrepreneur Thomas Elvin Edison in Menlo Park, New Jersey, the phonograph was an instrument for recording and playback of analog sounds. His earliest phonograph used cylinders consisting of an engraved audio recording that came in contact with a stylus that made contact and tracked the spiral groove, from one end to the other. The sound was amplified by a large horn. Edison was granted his first patent on February 1878.

 Plasma display   **1964**

A flat panel display, the plasma display is common to large TV displays and commercial monitors, although its concept has now been superseded by the LED, OLED and LCD displays. The plasma display operates by grouping many micro cells sandwiched between two glass panels that contain an inert mix of noble gasses. The gas in the cells is then electrically converted into a plasma that excites the phosphors in the gas, emitting light. Co-invented by University of Illinois researchers, H Gene Slottow, Donald Bitzer and graduate student Robert Wilson in 1964 as the 'monochrome plasma video display'.

 Postcodes (zip codes)   **1857**

The earliest system of postal districts was implemented in London and other large UK cities in 1857. In London, the system was further refined and expanded in 1917 to include numbered sub-areas, extending to other cities and large towns in 1934. The earlier district postcode systems were later incorporated into what is now the national postcode system.

 Postage stamps   **1680**

British merchant William Dockwra devised the world's first pre-paid postage system, the London Penny Post. The Penny Post was the world's first inter-city mail delivery service at a fixed price and formed the basis for many other provincial Penny Post services in the years that followed.

Communication – Audio-Visual

Rotary dial — 1891
A device mounted in or on a switchboard or telephone, the rotary dial was devised to send electrical pulses dependent and unique to the number dialed, known as pulse dialing. Invented by Almon Brown Strowger, he filed for a patent in 1891 and it was granted in November 1892.

Sign language — 1680
British intellectual George Dalgarno developed the world's first international linguistic sign language for deaf people. Dalgarno's method of sign language is used widely throughout Britain and the Commonwealth and still used in the USA.

SMS - text message — 1992
British mobile telecommunications engineer Neil Papworth was working for telecom company Vodafone, when he became the world's first person to send an SMS or text message in 1992. Papworth sent the text from his computer to Richard Jarvis, a field engineer, on his Orbital 901 mobile phone, with the message 'Happy Christmas'. Texting is now the single most popular method to stay in contact. According to British regulator OFCOM, 12-15-year-olds now send an average of 193 texts every week, with the average person sending 50 per week.

Stereo sound — 1931
Alan Dower Blumlein patented a technique for playing stereo sound with a single pick up and stylus. Later employed by EMI in 1933.

Telautograph — 1888
An analog predecessor to the contemporary fax machine, the telautograph transmitted electrical impulses logged by variable potentiometers from the transmitting location to stepping motors at the receiving station, to which a pen was attached. This enabled a signature or drawing to be transmitted and duplicated point to point. This invention is credited to Elisha Gray who was granted a US patent in 1888.

35

GB & USA: The Mothers of Invention

 ### Telegraph - electric   1837

Sir Charles Wheatstone & William Cooke filed a patent for the electric telegraph in 1837, which was granted the same year. The first fully functioning telegraph ran between West Drayton and Paddington railway stations in London, but the revolutionary invention's huge benefits were not capitalized. That was until 1845 on New Year's Day, when the electric telegraph helped seize murderer John Tawell. It caused a sensation and from this point on, Telegraph systems were soon spreading globally throughout the civilized world.

 ### Telegraph - printing   1846

The inventor of the printing telegraph, a variation of the electrical telegraph, was Royal Earl House from Rockland, Vermont. The device linked 2 identical 28 key piano style keyboards via 24 pairs of electric wires (one for each key). When a key was pressed at the transmitting or senders end, the corresponding letter would print out at the receiving end onto a paper strip, providing a clear readable text message for the receiver.

 ### Teleprompter   1950

A display device that prompts an announcer or presenter speaking via a visual aid is commonly known as a teleprompter. While working for 20th Century Fox film studios, the teleprompter was Invented in 1950 by Hubert Schlafly.

 ### Telephone   1876

Alexander Graham Bell applied for a patent for the telephone in 1876; later granted by the US patent office. British inventor, Bell patented his telephone model literally hours before a rival inventor. It evolved largely due to a discovery using a thin metal sheet that vibrated in an electromagnetic field. This generated an electrical waveform equivalent to the vibration. A demonstration of this invention was first publicly displayed in 1876 at the Centennial Exhibition in Philadelphia.

Communication – Audio-Visual

Television - cathode ray tube   1908

British engineer, Alan Archibald Campbell Swinton, published the world's first available article that proposed an electronic television using a cathode ray tube.

Television   1925

John Logie Baird submitted a patent for a mechanical television system in 1925. He later demonstrated his design in 1926; the first to transmit fully moving pictures. Unfortunately, his mechanical system was futile, with the advent of a competing alternative, offering a noticeable superior picture quality than Baird's system.

Television - transatlantic transmission   1928

British inventor John Logie Baird made the first transatlantic television transmission by land cable to an international radio transmitting and receiving base, using his electro mechanical system.

Television - colour   1928

John Logie Baird made the first colour television transmission using scanning discs at the receiving and transmitting terminations. He used three spirals of apertures with a different primary colour filter in each one. He installed three lights at the receiving termination point and used a commutator to oscillate the illumination. This was to be the world's first colour Television Transmission.

Ticker tape   1867

Edward A Calahan was an employee of the American Telegraph Company. He is cited for inventing and developing ticker tape as a means of conveying stock price information across telegraph wires. The device consisted of a strip of paper running through a machine known as a stock ticker, printing abbreviated codes for companies, prices and volume data. At the time, this method revolutionized the US stock market.

Transatlantic cable 1850

The world's first undersea cable laid in the English Channel was installed between England and France, followed by a transatlantic cable in 1857 led by British telegraph engineer, John Watkins Brett. He had a vision that telegraph cabling would create a new world of instantaneous communication. In 1847 the French government approved Brett's application to lay a submarine cable between Dover and Calais. Only a handful of calls were made before a fisherman struck and severed the cable, rendering it inoperable. In 1851, Brett founded the Submarine Company and by 1856 had raised enough finance to fund the Atlantic Telegraph Company, who laid the first transatlantic cable, but due to poor quality and installation techniques it had failed within the first few weeks of operation. Unfortunately, Brett died before witnessing the first successful cable installation later that year in 1866.

 Video recorder 1928

John Logie Baird filed the first patent for video recording, to record moving images on wax discs. His device was a by-product of his ground-breaking television invention and he named this new device Phonovision. It was constructed by the use of a large Nipkow disk attached mechanically to a standard 78rpm record cutting lathe. This produced a permanent record in 30-line analogue video resolution that could be played back through his television at 30-line quality. There was no further development or commercialisation.

 Video tape 1956

A means of recording, storing and playback of both sound and vision on magnetic tape, video tape was first invented and introduced by Charles Ginsburg and Ray Dolby in 1956.

 Voice mail 1973

The managing of telephone messages from a central data storage system, voice mail is typically stored on central data center servers, to recall if for instance, a caller cannot get through to the recipient. The inventor is Stephen J Boise in 1973 and commercialized in 1975.

 Whistles 1883

Joseph Hudson was a British toolmaker from Birmingham who converted his tiny washroom in his rented flat into a makeshift workshop. He worked on anything he could to supplement his meagre income, even repairing shoes at weekends. He crafted his first whistle in this workshop and later formed a company with his brother James, called Acme Whistles. It is still the world's most famous and largest producer of whistles today, still based in the Jewellery quarter in the City of Birmingham.

Criminology and crime

 Arsenic Test 1836

British chemist James Marsh developed the Marsh test. It is a highly accurate method for the detection of arsenic, particularly useful in forensic toxicology to detect arsenic as a poison. Arsenic used to be a highly successful and undetectable poison prior to the Marsh test and was almost untraceable in the body.

 DNA - computing 1994

Using DNA molecular biology and biochemistry rather than traditional silicon-based computing technology, DNA computing or molecular computing is a rapid changing interdisciplinary sector. The inventor of the principal of DNA computing is Leonard Adleman of the University of Southern California who pioneered this form of DNA data analysis in 1994.

Criminology and crime

DNA - database   1995

On April 10th, 1995, the world's first DNA database was launched in Britain and was cited as being the biggest breakthrough in the fight against crime since fingerprints, by the then Home Secretary Michael Howard.

DNA - fingerprinting   1984

After seven intensive years of research, Alan Jeffreys, a geneticist from Leicester University, developed a technique for distinguishing one person's DNA from another. It was on September 10th, 1984 when he suddenly realized differences and similarities in the DNA between his family members and one of his technicians at his laboratory. It was then that he realized the huge potential of his DNA profiling for use in police forensic work and soon worked with police on the murders of local girls Dawn Ashworth and Lynda Mann, both raped and murdered in Narborough, Leicestershire.

Tests proved that the DNA from both semen samples were from the same man and not from the prime suspect who had previously been arrested because he had the same blood group.

It was proved beyond doubt that this man was innocent. Jeffreys subsequently filed a patent for the use of DNA fingerprinting to identify individuals following real life tests over many years of research, concluding with tests on the rapist and murderer, Colin Pitchfork, proving beyond doubt that he was responsible for this heinous crime in Narborough, and a patent was subsequently granted in 1986. This technique has now been used to catch thousands of criminals worldwide through conclusive DNA evidence.

Face recognition   2008

A British joint venture between BAE and Omni Perception has improved suspect identification on CCTV footage and developed lip movement with speech pattern recognition technology. This new combination of speech and lip-reading recognition offers police and other law enforcement agencies a means to secure valuable evidence about a suspect's identity from CCTV, far quicker and more accurately than has ever been possible before.

Fingerprinting   1684

British scientist Nehemiah Grew is recognised as the first person to understanding and document the detail and relationship of fingerprints. Grew used a microscope to study and research the ridges found on palms and fingers and published his first paper on the subject in 1684.

Polygraph   1921

Traditionally known as a lie detector, the polygraph is an apparatus that measures and records numerous physiological indicators such as skin conductivity, blood pressure and heart beat, whilst the interrogator or questionee is being asked a series of questions. The responses should give the operator a degree of confidence if the subject is deceiving or lying. The polygraph was invented by John Augustus Larson, a medical student at the University of California and a police officer of Berkeley Police department, California in 1921.

Robert Melias   1987

Robert Melias, sentenced for 8 years in prison at Bristol Crown Court for Rape, notoriously became the first person in the world to be convicted of a crime using the evidence of DNA.

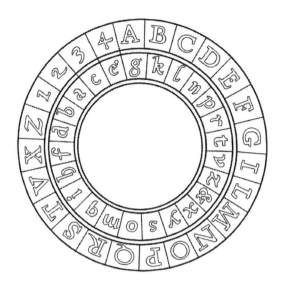

Cryptology

 Blind signature   1983

A form of digital signature where the message content is disguised prior to signing, a blind signature can be verified publicly against the original message. Examples of its use include digital cash processes and secure election systems. The blind signature was invented by David Chaum in 1983.

 Cipher - Bacons cipher   1605

Francis Bacon devised a technique to hide a secret message that was hidden in the body of text, rather than just a cipher. This method of encryption was known as Bacons Cipher or Baconian Cipher.

Cipher - Playfair Cipher   **1854**

Charles Wheatstone invented the world's first manual symmetric encryption cipher but bore the name 'Playfair Cipher' because Lord Playfair was the promoter of this unique device. The cipher encrypted pairs of letters, rather than earlier ciphers that encrypted single letters. This made its messages much harder to break, since the method of frequency analysis used on simpler ciphers did not work well. The device was used in the Second Boer War, WWI and WWII because it was quick to set up and use and fairly reliable.

 RSA cipher   **1973**

Clifford Cocks is a mathematician and cryptographer, who, while based at GCHQ in Cheltenham (GB), discovered the encryption algorithm now commonly used and known as RSA. His algorithm was created more than three years prior to being independently developed by Rivest, Shamir and Adleman at MIT in the USA during 1977. At the time he was not recognised for his creation because working for GCHQ meant that his work was classified information and therefore not available in the public domain.

 Stream cipher   **1917**

Also known as a state cipher, invented by Gilbert Sandford Vernam at Bell Laboratories, the stream cipher is a symmetric key cipher where a pseudorandom cipher bit stream is combined with plain text bits. The plain text bits are encrypted one at a time in a stream cipher, transforming successive digits during the encryption.

 Wheel cipher   **1795**

The wheel cypher, formerly known as the Jefferson disk, was a basic cipher system designed to encrypt messages as a deterrent for codebreaking. Invented by Thomas Jefferson, the device comprised of 26 wheels, with a complete set of alphabet letters on each wheel, randomly arranged and different on each disk. The principle was that the coder used exactly the same wheel as the decoder, hence written coded messages could be unscrambled securely.

Digital and computing world

ATM (Cash machine) PIN 1966

James Goodfellow patented the Personal Identification Number (PIN) technology and is generally acknowledged as one of the most significant co-inventors of the ATM (Automatic Teller Machine). Goodfellow was a British Engineer and given the project to develop an automatic cash dispenser in 1965. His system allowed a machine-readable card to be used. The cards were encrypted for security through the use of magnetic strips allied to a four-digit PIN number, accessed by the ATM keypad.

ATM (Cash machine) 1967

The world's first ATM or cash machine was established at a branch of Barclays Bank in Enfield, North London. John Shepherd Barron was the chief inventor and worked for the world's largest bank note printing firm, De La Rue. The first person in the world to use the ATM was comedy actor, Reg Varney, star of the TV sitcom 'On the buses. At the time, the ATMs were known as DACs (De La Rue Automatic Cash System). Shepherd Barron was the General Manager for De La Rue Instruments who made the world's first ATM's.

ATM - concept   1939

Now known as a computerized telecommunication machine, providing clients associated with a financial or banking institution, access to their account details. Clients also have the facility to withdraw or deposit cash by using a PIN secured bank card for security. Although before the golden age of global computerization, inventor Luther George Simjian, devised the basic concept of a 'hole in the wall' to enable customers of banks to access their funds out of banking hours and at the time he registered many patents to secure the concept. Though, it would be many years before the communication infrastructure, computing and base concept could be refined and developed to a standard of security and practicality acceptable to the public and financial institutions.

 ARG - electrically powered analogue computer   1906

Arthur Pollen was a British naval affairs writer and through his huge naval experience recognised the need for a computer-based fire control system on war ships, due to the wholly inaccurate method of manual firing of the day. He set about designing a suitable system in the early 1900s and later produced the first version of what is now recognised as the first computerized fire control system. It took account of the tide and its effect on the ship by using a tidal analyzer.

Later in more advanced versions; Pollen integrated a gyroscope to take account of the ships yaw and used a graph plotter to capture the target's position and relative motion. For the first time ever, the British Navy had automatic assistance in range and firing, producing startlingly accurate results.

ARM - universal mobile device processor   1978

Mobile phone processor designer, Arm Holdings based in Cambridge, has grown from making one of the world's first home and school computers, the Acorn, to being one of the world's most important semi-conductor designers. The ARM processor provides the heart to more than 95% of smart-phones and tablets globally; including iPad, iPhone, Microsoft, Samsung, Sony and LG. Significantly, ARM holdings is not a manufacturer, but a designer and licensor of its power processors. In 2008, the ten billionth ARM based processor was shipped.

Bar codes commercial application **1970**

Bar coding was first used commercially in 1966, but it was not until British company Plessey developed a universally acceptable system, that its use became widespread and internationally accepted. The first company to universally adopt this system of stock control was the British food retailer J Sainsbury, used for every item of stock in their portfolio.

 BASIC **1963**

A family of high-level programming languages, BASIC was invented by Thomas Eugene Kurtz and John George Kemeny at Dartford College, New Hampshire in 1963. The aim of the invention was to produce a programming protocol that was accessible to non-science students. The BASIC language and its variants became prevalent on personal computers and desk tops from the late 70's.

 Biometrics - iris recognition **1949**

British ophthalmologist James Doggart wrote 'just as every human has different fingerprints, so does the minute architecture of the iris exhibit variations in every subject examined. Its features represent a series of variable factors whose conceivable permutations and combinations are almost infinite'. This was the first time a scientific paper had been written and published on this subject.

Biometrics - iris recognition algorithms **1994**

British ophthalmologist John Daugman developed and patented the first algorithms to recognize the individual iris of a human eye. Daugman's algorithms became extensively licensed by a series of different commercial organisations. Today, after many improvements, Daugman's algorithms remain the foundation of all commercial employment of iris recognition technology.

Calculator - pocket 1972

The world's first 'slimline' pocket calculator was launched in 1972 at the huge price of £79.95 Plus VAT. It was designed by British inventor and Scientist, Clive Sinclair and manufactured by Sinclair Radionics. In 1973 a new version was launched called the Sinclair Executive Memory. Sinclair went on to make many more breakthroughs, such as the ZX Spectrum, one of the first hand held computers and the Sinclair C5 3-wheel electric transporter.

CMOS 1994

The abbreviated name of Complementary Metal-Oxide Semi-Conductor is CMOS. It is an image sensor that is made up of an integrated circuit comprising an array of pixel sensors. There is a photo sensor in each pixel and an active amplifier. Beginning at the same position, each pixel converts light into electrons by using the CMOS method. CMOS image sensors can be found in digital SLR cameras, video cameras, automotive safety systems, embedded web-cams, toys and video games. The CMOS was invented and developed by the renowned engineer and physicist, Eric Fossum whilst working at NASA's Jet Propulsion Laboratory in Pasadena, California. A patent was granted to Fossum on November 28th, 1995.

Computer programme - first programmer 1843

British mathematician Ada Lovelace is recognised as being the world's first computer programmer, by writing instructions to drive the first proper program on one of inventor Charles Babbage's early computers, the Analytical Engine. Lovelace wrote and published a paper on the subject of programming and spent many years perfecting and refining her newfound art.

Computer - world's first   1820

Charles Babbage was a renowned philosopher, mathematician, engineer and inventor. Babbage is credited with creating the world's first mechanical computer, the 'Difference Engine'. Babbage designed this machine to automatically calculate polynomial functions. The name 'Difference' derives from the system of divided differences; a method to tabulate functions using a set of polynomial coefficients, both trigonometric and logarithmic functions.

Computer - first stored programme   1948

The SSEM was developed and built by British computer scientists Frederic C Williams, Tom Kilburn and Geoff Tootill, all from the Victoria University of Manchester. The Manchester Small Scale Experimental Machine (SSEM) was not intended to be a mass market computer, instead designed as an early test bed for the 'Williams tube', a primitive form of computer memory. The SSEM was the first working machine to enclose all the fundamental elements of a modern computer.

Computer - laptop   1979

British industrial designer, Bill Moggridge, designed the Grid Compass. It was the world's first commercial laptop computer with clamshell design and flat panel screen. Moggridge is attributed to designing a 'computer in a suitcase' and when it was released for sale, it had the staggering price of £5,091. The Grid Compass had a magnesium case with a yellow on black keyboard and was eventually used by the US military, later having the distinction of being used on the Space Shuttle.

Computer - portable 1981

Adam Osbourne, a British computer scientist is credited with developing the world's first commercial portable microcomputer, not to be confused with a laptop. US industrial designer Lee Felenstein designed the Kaypro aesthetics. Osbourne planned to break the high price of computers. The KayPro design was loosely based on the Xerox Note Taker. It was intended to be both robust and portable, clad in an ABS (Acrylonitrile butadiene styrene) case and hidden handle. Unfortunately, it was about the weight and size of an accordion and had the looks of a WW1 telecommunication radio. At its peak, the Osbourne 1 sold 10,000 units a month.

 ### Computer - handheld 1984

British computing pioneers PSION developed the Psion Organizer. The Psion Organizer is regarded as being the world's first handheld computer and was launched as the PSION 1, under the brand name PSION Organizer. It was the world's first useable PDA (Portable Digital Assistant) and was characterized by its sliding keyboard cover and LCD display. It weighed 225 grams and measured 142 x 78 x 29.3 mm, equipped with fully static RAM with impressive battery life for its time. It was originally priced at £99. In 1986, PSION introduced an improved version called the PSION 2.

 ### Computer - electronic programmable 1943

Tommy Flowers was an ingenious British Post office engineer by profession. But it was the war effort and great need for powerful encryption of coded communications from the German Lorenz Cipher that gave Flowers the inspiration to create the world's first electronic programmable computer – The Colossus. Far more complex and powerful than the more famous Enigma, it was the world's first truly programmable electronic computer, and used more than 1500 valves or vacuum tubes as they were then known. Unfortunately, such was the secrecy and the advanced power and technology of the Colossus, that it was only ever used for Military encryption and remained a secret until long after the Second World War was over. Unfortunately, every Colossus ever made was dismantled.

Diode - rectifier tube radio Valve   1904

Sir John Ambrose Fleming was a renowned scientist who developed the 'Fleming Valve' or oscillation Valve. It was the world's first thermionic or diode valve, effectively the electronic equivalent to a one-way water valve. It was the first thermionic diode and was used extensively in later years in radio receivers and also as a low current rectifier in power supply units.

 Ethernet   1975

The ethernet is a networking technology that is frame based, used primarily for local area networks (LAN's). Its name originates from ether or the physical concept of it connecting to a network. The inventor Robert Metcalfe was working for Xerox when he created the ethernet in 1975.

Fax machine   1843

British engineer and inventor Alexander Bain was the first person in the world to file a patent for the facsimile machine (FAX), but with the exceptions of a few improvements in the years ahead did not develop it. This was an opportunity for Frederick Bakewell who beat Bain with a demonstrable model of a fax machine that he called the 'image telegraph'.

Firewall   1988

A collection of security procedures designed to prevent illegal and intrusive electronic access to a network of computers. A firewall is incorporated into almost all domestic and commercial computer systems. Its inventors are AT&T Bell Labs employees, Steven M Bellovin and William Cheswick who created the firewall whilst working on packet filtering research.

Graphic user interface   **1981**

GUI, short for graphic user interface, utilizes menus, icons and windows to prompt instructions such as moving files, opening files, deleting files and many other general commands, typically by using a mouse or a screen pen. The GUI was invented in 1981 by engineers from Xerox PARC, Douglas Engelbart and Alan Kay.

 Hard disk drive - mainframes   **1955**

Typically, a non-volatile storage device that collects data digitally on rapidly rotating magnetic disks, the hard drive for commercial computing systems was invented by Reynold Johnson in 1955.

 Hard drives - for personal computers   **1997**

British Professor Stuart Parkin developed high capacity computer data readers. These devices allowed more information to be stored on each disc platter than previous incarnations and this invention has now laid firm foundations in data storage protocol and development, enabling the storage needed for huge organisations such as Amazon, Google, Facebook and other online services.

 HTTP URLs and HTML   **1989**

Famed for his development of the World Wide Web, British computer scientist and Oxford educated Tim Berners-Lee also simultaneously developed URLs, HTTP and HTML, later refined as web technology rapidly developed.

 Hypertext   **1965**

Often referring to text on a computer that will lead one user to another on demand. Hypertext overcomes many of the limitations in the written word. Hypertext was invented by Ted Nelson in 1965, who then went on to develop the Hypertext Editing System in 1968.

Digital and Computing World

Integrated circuit or microchip　　　　　　1952

G.W.A. Dummer was an electronics engineer from Hull, who was the first person to create and construct a prototype of the Integrated Circuit in 1952, better known as the Microchip. Through his work at the Telecommunications Research Centre, he believed that it was feasible to build multiple circuits on and into silicon or a similar substance. While at a conference in Washington DC, he presented his work. Six years later, Jack Kilby of Texas Instruments was awarded a patent for essentially the same design raising many questions about the legitimacy of the award. Foyle is known as the 'father' or 'prophet of the integrated circuit'.

 Internet　　　　　　1983

Often mistaken as WWW being one of the same, when in fact both the world wide web and the internet, although inter-related, are two completely separate applications. The world wide web, invented by British computer engineer Sir Tim Berners-Lee, was invented in the early 1990's, whereas the internet was created at least seven years prior. The internet is a global network of interconnected computers and data storage hubs that utilize the standardized Internet protocol (IP) suite, that serves billions of users globally. The internet is effectively a web of networks all linked by fiber optics, wireless and cable. The first electronic network was established in 1969 between Leonard Kleinrocks UCLA lab and Douglas Engelbart's Stanford Research Institutes lab. Later in 1973, While working at US Department of Defense's ARPANET facility, Vinton Cerf and Bob Kahn co-invented internet protocol (IP) and transmission control protocol (TCP). Then later in 1983, the world's first TCP-IP wide area network was operational and thus spawned the beginning of the Internet as we know it today.

iPod® Origins　　　　　　1979

British electronics engineer Kane Kramer, patented the revolutionary IXI player, a device that stored music on a digital chip. Holding just three and a half minutes of music, barely enough for a full single. His patents ran out in 1988 and because he couldn't raise the £60,000 needed to renew them, the Apple® iPod was revealed to the world in 2001.

iPod® 2001

Designed by British born Apple chief designer Sir Jonathan Ive, who is one of the senior directors of the giant computer company; Apple®. The iPod® has since sold more than 170 million units. The Daily Telegraph recognised Jonathan Ive in 2008, as being the most influential Briton living in America today.

 Microprocessor 1971

An integrated computer chip designed to process instructions and link up with external devices, additionally centrally processing most of the key functions within a computer but built on a single silicon chip. The Intel 4004 was co-invented by Federico Faggin, Stanley Mazor and Ted Hoff whilst working for a calculator company called Busicom.

 Packet switching 1965

Donald Davies is an inventor who developed the Internet protocol for sending and receiving information in parcels or packets of information. Now taken for granted, the way information or data is packed and transmitted on both the Internet and many mobile phone systems globally is the de-facto standard for all operating systems.

 Programming languages 1936

Programming languages are typically machine-readable artificial languages that can be employed to create programs that control the behaviour of a machine by expressing algorithms accurately. Although computer programming in its most basic form happened in the last part of the 19th century in Britain, this most modern form of calculus, created by Stephen Cole Kleene and Alonzo Church in the early 1930's has now developed into a useful tool in computerized problem solving and as an aid in the development of software languages.

Qwerty keyboard   1874
The ubiquitous QWERTY keyboard that we all use on our computers and handheld devices today, was originally devised and invented by Christopher Sholes, who was granted a patent in 1874. The name originates from the first six letters from the left of the keyboard.

 ### RAM - Random Access Memory   1918
William Eccles and F.W. Jordan were British physicists who developed the 'flip-flop' circuit in 1918. This methodology became the foundation of electronic memory in computers still used worldwide today.

 ### Virtual reality   1968
A technology that allows the user interaction with a computer-simulated environment, virtual reality are primarily visual experiences with audio added. Viewing is normally on computer screens or through stereoscopic virtual reality glasses or headset. The inventor of the world's first computer generated virtual reality was Ivan Sutherland aided by his student Bob Sproull using a head mounted display (HMD).

 ### Web browser   1990
British computer scientist Tim Berners-Lee invented the world's first web browser. It was then called WorldWideWeb (all one word) but was later renamed NEXUS. Berners-Lee used a neXT computer and pioneered the use and terminology of 'hypertext' for information sharing. He was also responsible for creating the world's first web server. Since his creation, the web and web browsers are seen as one by most users.

Wireless local area network 1970

The linking of two or more computers employing OFDM modulation or spread spectrum, a wireless local area network enables communication between devices with in a limited area. The first example of this network type was invented by a team of researchers at the University of Hawaii, led by Norman Abramson, who devised and operated the world's first computer network, using low cost CB or ham-type radios, then called ALOHAnet, deployed over 4 islands.

 WWW - World Wide Web 1991

The World Wide Web was developed and invented by computer scientist Sir Tim Berners-Lee. He created his first server in 1990 and on August 6th, 1991, his web went live with its first page describing how to search and set up a web site. Berners-Lee took no fee for his revolutionary invention, which he gave to the world completely free. This is not to be confused with the Internet, which is a network of linked computers

 Zip - file format 1989

Invented and developed by Phil Katz whilst working for PKZIP, the zip file format is a file and data compression archiver, particularly useful for folders that contain more than one file.

Discovery and establishment of the USA

 **America**   **1170**

British Prince Madog Ab Owain from Wales landed in the present-day town of Mobile, Alabama. Almost 350 years before Christopher Columbus landed in the Caribbean in 1492 and long before the first documented European, Spanish Conquistador Juan Prince de Leon arrived on the mainland of what is now known as the USA on April 2nd, 1513.

There is no doubt that Great Britain played a huge part in establishing the USA, the world's most formidable super power in the guise we see today. The USA was created when established British colonies in North America declared independence from what was then known as the Kingdom of Great Britain. The motivation that led to this break-up was concerning taxation and what was seen as continually rising duty for no apparent benefit to the people of the American colonies. The War of Independence lasted from 1775 to 1783 with the American colonial forces, aided by French, Dutch and Spanish forces, claiming ultimate victory. The ironic reality, however controversial, remains though that the victors were British subjects until that day of Independence in 1783.

American overseas territories

Any geographical location that is under control of the United States Federal Government outside the main states of the USA is considered American overseas territory. A United States territory (but not limited to these areas) includes a clearly defined geographic area and refers to an area of land, sea or air under jurisdiction of United States federal governmental authority. The degree of territory is all the area owned by and under the protectorate of, the United States of America Federal Government (that may comprise areas some distance from the nation) for governmental and other reasons.

The US overseas territories are:

- American Samoa
- Bajo Nuevo Bank
- Baker Island
- Guam
- Howland Island
- Jarvis Island
- Johnston Atoll
- Kingman Reef
- Midway Islands
- Navassa Island
- Northern Mariana Islands
- Palmyra Atoll
- Puerto Rico
- Serranilla Bank
- US Virgin Islands
- Wake Island

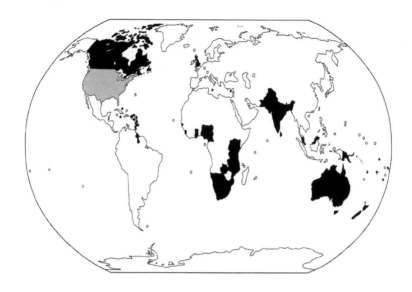

British Commonwealth countries - past and more recent

British overseas territories

More than 14 former colonies remain under British rule, although the word 'colonies' is officially no longer used to describe these. Without exception all 14 of the British overseas territories are small islands with a small population and the majority are in extremely remote areas of the world. All territories have a degree of self-governing autonomy with the United Kingdom taking responsibility for defence and external affairs.

The British overseas territories are:

- Anguilla
- Bermuda
- British Antarctic territory
- British Indian Ocean Territory
- British Virgin Islands
- Cayman Islands
- Falkland Islands
- Gibraltar
- Montserrat
- Pitcairn Islands
- St Helena, Ascension and Tristan da Cunha
- South Georgia and the South Sandwich Islands
- Sovereign base areas of Akrotiri and Dhekelia
- Turks and Caicos Islands

British Crown dependencies

- Guernsey
- Isle of Man
- Jersey

The Commonwealth   1583

The original British Empire, between 1583 and 1783, developed when Walter Raleigh and Humphrey Gilbert were granted permission by Queen Elizabeth I to establish colonies in North America.

The second phase of the British Empire between 1784 and 1815 majored on controlling governments and infrastructure of the colonies and the increase of trade. India was a good example, where in the latter part of the 18th century, the ruling Mughals relinquished power to the British.

The third phase of the British Empire between 1815 and 1914 was known as the Imperial Century. It was in this decade that Britain controlled more than 10 million square miles of the World, inhabited by more than 400 million people. Britain was such a global force, that even economies that it did not govern, such as China, were still controlled from London.

At the start of the 20th century, there was a latent wave of countries wishing to become independent from the direct rule of Britain but liked the idea of still remaining within the family of colonial Britain. In fact, as far back as 1814, during a visit to Australia, Lord Rosebery first described the evolving nature of the British Empire by calling it a 'Commonwealth of Nations'. Then in 1921, the term 'British Commonwealth' was first officially used by statute, to replace 'British Empire' in the wording of the 'Anglo-Irish Treaty' of the same year.

In 1926 at the Imperial Conference, Britain agreed that it and all its dominions were of equal status in every aspect, and that all were united in their allegiance to the Crown. From this point on, this great group became known as 'The British Commonwealth Nations'.

After WWII, the former British Empire was slowly dismantled, leaving only 14 wholly dependent territories within the family of the British Commonwealth, and in 1949, the term 'British' was removed from its name, leaving 'The Commonwealth' to truly reflect its changing nature and role in the modern world.

Today, The Commonwealth includes some countries that are newly formed by modern standards or which were not previously part of the old British Empire. The current membership of Commonwealth countries (2014), totals 53 nations in all. The membership includes countries spanning the Americas, Africa, Asia, the Pacific and Europe. The countries range from some of the largest countries in the world to some of the smallest countries. They also vary from some of the poorest to the richest countries. All members of the Commonwealth have to pledge to its principles and values as written in 'The Commonwealth Charter'.

Regardless of economic status or size, all member countries have an equal say and vote and this protocol, ensuring that even the smallest countries have an equal say in developing the Commonwealth. The leaders of each member country meet up bi-annually to discuss the issues affecting both the wider world and The Commonwealth, with the one aim of shaping its priorities and policies.

English language

English is spoken as a main language by a fifth of the Earth's population. The story of English as a language is one of struggle and survival against all odds, over many centuries encapsulating its development and evolution. The English language has been shaped and influenced over the years by Romans and Italian scholars who introduced Latin; the German influence derived from their barbaric tribes; Norse influence from the Viking invasions and Normans from France who introduced many French sayings that we now take for granted. With all these external pressures, the English language has had to exert a huge amount of resistance. More people, simply by virtue of China's huge population, speak Chinese Mandarin. However, English is now central to science, business, international trade, Internet, international air traffic control, world politics and diplomacy worldwide.

 Australia - formation   **1770-1788**

Captain James Cook was a British explorer, cartographer and navigator in the Royal Navy. He produced detailed maps of Newfoundland before making three voyages to the Pacific Ocean, in which he made his first contact with the Hawaiian Islands and the eastern coastline of Australia. He also recorded the world's first circumnavigation of what is now New Zealand. As he advanced his voyages of discovery, he surveyed, mapped and started to name features. He recorded and started to name islands and coastlines of these new far-flung lands for the first time

Cook subsequently charted the Australian east coast in his ship HM Bark Endeavour. Under instruction from King George III of England, Cook claimed the Australian east coast in 1770. He named Eastern Australia as 'New South Wales'. Cook was sadly killed during his third Pacific trip to Hawaii during a fight with locals. His legacy of geography, cartography and scientific knowledge was to influence his successors into the 20th century. Then in 1788, Captain Arthur Phillips and his first fleet, with 11 ships and 1350 people, arrived at what is now known as Botany Bay, moving quickly on to Port Jackson because Botany Bay was considered unsuitable for settlement. Captain Phillips had instructions to establish the first British colony in Australia in the Sydney Cove area, now the harbor of one of Australia's greatest cities; Sydney.

The Commonwealth members 2014

Africa (17)

- Botswana
- Cameroon
- Ghana
- Kenya
- Lesotho
- Malawi
- Mauritius
- Mozambique
- Namibia
- Nigeria
- Rwanda
- Seychelles
- Sierra Leone
- South Africa
- Swaziland
- Uganda
- United Republic of Tanzania

Caribbean and Americas (13)

- Antigua and Barbuda
- Bahamas
- Barbados
- Belize
- Canada
- Dominica
- Grenada
- Guyana
- Jamaica
- Saint Lucia
- St Kitts and Nevis
- St Vincent and The Grenadines
- Trinidad and Tobago

Asia (7)

- Bangladesh
- Brunei Darussalam
- India
- Malaysia
- Pakistan
- Singapore
- Sri Lanka

Europe (3)

- Cyprus
- Malta
- United Kingdom

Pacific (11)

- Australia
- Fiji
- Kiribati
- Nauru
- New Zealand
- Papua New Guinea
- Samoa
- Solomon Islands
- Tonga
- Tuvalu
- Vanuatu

New Zealand - discovery and settlement   **1769**

The establishment of people in New Zealand dates back at least 700 years when Polynesians settled and started forming what is now recognised as the Maori culture.

Then in the early 17th century, the Dutch, fresh from victory over the Spanish, grabbed most of the East Indies from the Portuguese and so keen to develop the great continent of the south, that many people believed lay between Cape Horn and Australia, commenced a voyage of discovery. They first discovered Tasmania, named after the Dutch captain Tasman and encountered fierce opposition from the local Maori warriors. He quickly cast off and headed north; unaware that he was just a few hundred miles off the 'great southern continent'.

It was left to British naval commander and explorer James Cook to finally discover what is now known as New Zealand in 1769. Then in 1840 after many years of negotiation between the British government and the Native Maoris, the Treaty of Waitangi was signed welcoming New Zealand into the British Empire, giving Maoris the same rights as British subjects.

"One doesn't discover new lands without consenting to lose sight of the shore for a very long time."

Andre Gide

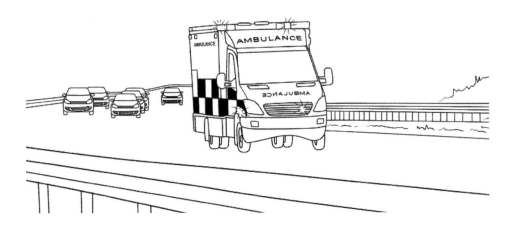

Emergency services & devices

 Fire brigade   **1824**

British Fire Master James Braidwood established the world's first municipal fire brigade in Edinburgh and is recognised as being the founder and developer of the modern metropolitan fire service.

 Police - world's oldest police force   **1829**

The Metropolitan Police Act of 1829 formed the basis for London's Metropolitan police force that was founded in the same year. The 'Met' as it is now affectionately known, was a merger of the Bow Street Horse Patrol and the River Thames Marine Police Force dating back to 1805. Still in operation today, the Metropolitan Police force is the world's oldest continuously operating police force in the world.

Fire extinguisher - modern portable　　　　　　1816

British inventor George William Manby developed the world's first pressurized cylinder fire extinguisher. Manby witnessed firemen struggling to extinguish a blaze at a house fire in Edinburgh. This prompted him to design a portable copper made cylinder holding about four gallons of potassium carbonate, dispersed by air pre-compressed with a regulator to control the output when operated.

Lifeboat - world's first　　　　　　1774

British engineer Lionel Lukin submitted a patent for the world's first non-submersible lifeboat. He modified and patented a 20ft Norwegian yawl and fitted it with watertight cork filled chambers for extra buoyancy and then fitted a cast iron keel to keep the boat upright.

Lifeboat station　　　　　　1776

The world's first lifeboat station opened in Bamburgh, Northumberland, with a LUKIN lifeboat. Its claim to fame is that it was the scene of the world's first lifeboat launch. The original lifeboat house has since been converted into a quaint self-catering cottage.

Taser　　　　　　1969

Typically, a weapon designed to produce an electric shock, the Taser uses electro-muscular disruption (EMD) technology to incapacitate the receiver. It is gradually becoming the first choice in law enforcement rather than the conventional gun or stun gun. It was invented and developed in 1969 by NASA researcher, Jack Cover.

Water sprinkler 1723

The world's first water sprinkler system was designed and developed by chemist, Ambrose Godfrey. The device consisted of a barrel of water mixed with chemical dry powder of Godfrey's design. Included in the barrel was a sealed tin containing gunpowder with an external fuse. The patent filed by Godfrey and granted in 1723, informs that the fuse was lit by the flames of the fire, then the gunpowder exploded, splitting the barrel, thereby splashing the water and chemical contents around to extinguish the flames. It proved to be completely successful in at least one test.

"Emergency services… a good day is when everybody gets to go home…"

Anon

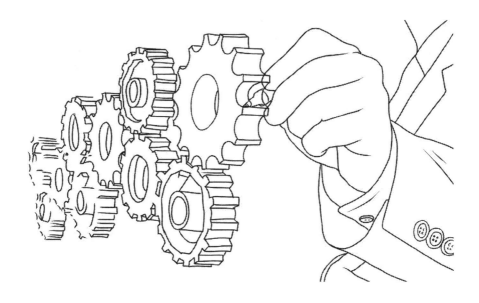

Engineering

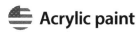 **Acrylic paint**   **1947**

A fast-drying paint that comprises a pigment that is suspended in an acrylic polymer emulsion, acrylic paint was invented by Leonard Bocour and Sam Golden under the brand name of 'Magna Paint' in 1947.

 Adjustable spanner or wrench   **1830**

Edward Budding was an engineer from Stroud in Gloucestershire who invented the world's first adjustable spanner. Inventor and British Engineer Richard Clyburn later improved the design in 1842.

 Aerosol paint   **1949**

A category of paint that is delivered via a sealed pressurized container, the aerosol releases a fine spray when the user presses a valve lever or button, normally positioned on the top of the container. Based on the same method of delivery as spray insecticides and deodorant's, aerosol paint was invented in 1949 by Ed Seymour from Sycamore, Illinois.

Air conditioning 1902

The process of de-humidification and cooling of internal air in a room or space for thermal comfort is known as air conditioning. Willis Carrier invented and manufactured the first mechanical air conditioning device. Carrier's invention paved the way to what is now a huge industry that created profound cultural and socio-economic change around the world.

 ### Aquarium - design and construction 1853

Phillip Gosse was a naturalist, who conceived and developed the world's first large-scale aquarium. His creation was not only used to display aquatic species, but he also designed the huge tank to be used for plants and reptiles. Thus, in the process, Gosse opened the world's first public display aquarium, completed and open to the public in 1853 at Regent's Park, London.

 ### Ball bearing 1794

Phillip Vaughan took out the first patent on ball bearings in this year. Up until this pioneering invention, moving parts on machinery tended to wear out very quickly as they rubbed against each other, but Vaughan came up with the notion of using solid steel balls secured in a groove to reduce friction and take the stress and strain out of moving parts. This great invention is now used in a plethora of products from skateboards to wheels on cars and from lorries to drive shafts on petrol and electric vehicles.

 ### Bottle cap – crown cork 1892

The first form of a true bottle cap was the crown cork cap, invented and patented by William painter from Baltimore, Maryland. It comprised metal flanges bent over the top rim of a sealed bottle to compress the liquid inside and was the forerunner to the effective and hygienic caps typically used on beer bottles globally today.

Engineering

 Bank notes - world's largest manufacturer   19th century

British company De La Rue has been established for more than 200 years and is the world's largest commercial manufacturer of Bank Notes. The British company is the official bank note producer for more than 150 nations worldwide and is also responsible for officially producing the passports for more than 80 countries globally.

 Bridge - box girder   1848

British engineer Robert Stephenson built the world's first box girder bridge, a railway bridge built between Llandudno and the town of Conwy in Wales. The Conwy wrought iron bridge was built to a design by engineer and bridge designer William Fairbairn and constructed by British inventor and engineer Robert Stephenson. The bridge was completed in 1848 but not officially opened until 1849. Because this was the world's first ever bridge to be built in this form, the designers of the concept, Fairbairn and Stephenson, needed to carry out intensive proving of prototypes to demonstrate that it could safely carry the huge weight of railway locomotives. Aside from reinforcing columns being introduced under the bridge in later years due to the advent of even heavier and more frequent locomotives, the bridge remains intact since its original construction.

 Bridge - concrete   1877

The world's first concrete bridge was built in Seaton, Devon, over the River Axe. British engineer, Philip Brannan built the triple span concrete arch structure in 1877, with the middle arch spanning 50 feet. The bridge still stands to this day.

 Bridge - iron   1779

British engineer Abraham Darby was a pioneer in the process of smelting local iron ore by using coke from local coal. He knew that industrial development was being held back in his area by the lack of a bridge over the River Severn. The main issue was that the bridge had to be single span due to the huge amount of barge traffic on the river. An Iron bridge was originally suggested and designed by Thomas Prichard, an architect from

Shrewsbury, but he died just as work commenced. Abraham Darby III who had the resource and the knowledge to build such a ground-breaking bridge design, then adopted the project. The work commenced and after more than 380 tons of iron was transported from Darby's smelting works to build the 33yard (30m) span bridge, it officially opened on New Year's Day 1781.

Bridge - railway — 1727

British engineer Ralph Wood built the world's first railway bridge near Stanley, County Durham, known as the Causey Arch; it is the world's oldest surviving single arch railway bridge. The Causey Arch Bridge was funded by the 'Grand Allies', a local consortium of coal-owners and designed and built by local stonemason Ralph Wood.

Documentation from archives confirm that the bridge cost £12,000 to build and was used to take coal from the pits to the River Tyne and then back via horse drawn wagons on a double track. It was the longest single span arch bridge in the country when it was built. The Causey Arch Bridge was reinforced and restored in the 1980s and now forms part of a series of scenic walking routes in the area around the Causey Burn.

Bridge - suspension — 1801

Characteristically, a suspension bridge is a structure in which the load bearing deck that spans between two points, is suspended from vertical hangers or cables, carrying the weight of the deck below. It is believed that primitive versions made entirely from twine, were developed more than 3000 years ago in countries such as China and India, usually spanning a gorge or river. Although the first modern suspension bridge, capable of carrying huge loads from pedestrians and small horse and carts below is credited to inventor James Finley of Uniontown, Pennsylvania. His design employed two vertical towers at either end of the bridge, that raised two curved iron cables from which iron hangers were attached top and bottom. His design, albeit improved with the use of stronger steel cables now, still uses the same principle throughout the world.

Engineering

 Bridge - tilting   **2001**

British architects Wilkinson Eyre designed the world's first tilting bridge. The Gateshead Millennium Bridge is designed for cyclists and pedestrians only and traverses the River Tyne between the quayside of Newcastle-upon-Tyne and Gateshead's Southbank Quays quarter. The bridge won a design award and is sometimes referred to as the 'blinking eye bridge' due to its tilting method and shape.

 Bulldozer   **1923**

Co-invented by draftsman J Earl McLeod and farmer James Cummings from Morrow Ville, Kansas, the bulldozer was a continuous tracked tractor built with a heavy metal bucket or plate on the front. This machine was designed to move large masses of sand, rubble or soil during construction work.

 Burglar alarm   **1853**

Consisting of sensors, linked to a central control box or unit, either hardwired or today often linked by radio frequency wireless methods, a burglar alarm is used as a preventative measure that will sound off an audible alarm or send a signal direct to security services to attend the disturbance. The burglar alarm was originally invented by Harvard educated Dr William Channing from Boston, and he was granted a patent on June 21st, 1853.

 Carbon fiber   **1958**

A material comprising ultra-thin fibers that are composed mainly of carbon atoms, carbon fiber was invented by Dr Roger Bacon of the Union Carbide Parma Technical Center in Cleveland, Ohio.

 Caterpillar tracks - patent   **1770**

Richard Edgeworth was an inventor and writer, born in Bath. Among his many inventions, Edgeworth filed a patent for what is now commonly known as caterpillar tracks. He submitted a design for a steam engine that ran on an 'endless railway system'. Flat steel planks were attached to the main wheels of the steam engine to ensure operation on soft ground, but in reality, Edgeworth's system was not successful until 1846, when James Boydell submitted a more feasible system.

Caterpillar tracks - first commercial use   1903
The world's first vehicle to utilize caterpillar tracks was named a 'crawler tractor' and was built by David Roberts of Ruston Hornsby & Sons.

Circular saw   1813
Essentially a circular steel disk with saw teeth on the outer edge, powered by a machine that is motor driven, the circular saw was invented by Tabitha Babbit, first used in a saw mill in 1813. The device can now be either hand-held or table mounted.

Diesel engine - pre-diesel patent   1890
Herbert Akroyd-Stuart was a British inventor and mechanical Engineer noted for his 'Hot Bulb Engine' or heavy oil engine creation. He patented his compression ignition engine two years before German Rudolf Diesel. Akroyd-Stuart built his initial prototypes in 1896 and filed his first patent in 1890. His heavy oil engine was far more advanced than Diesel's engine. Furthermore, it was two years ahead in its submission and patent awards.

Elevator brake   1852
Used a key component of an elevator or lift, Elisha Graves Otis invented the world's first safety brake that prevents an elevator from dropping out of control between floors inside a building.

Fire sprinkler – automated   1874
The original automated fire sprinkler discharged water when the heat of a fire was detected at a pre-set temperature. This invention is credited to New Haven, Connecticut resident Henry S Parmalee when he successfully installed the first closed head automated fire sprinkler in 1874.

Engineering

 Fluorescent lamp – compact   **1936**

Invented and developed by George Inman while working for Genral Electric. The fluorescent lamp was designed to replace standard incandescent light bulbs. Some of the compact fluorescent lamps were interchangeable with normal bulbs, while others had bespoke contact points. Inman discovered early on that a fluorescent light would use 70% less power than a conventional light bulb of the same wattage and have a longer working life. He was awarded a patent for the design principle in 1936, although due to manufacturing process difficulties, his invention did not appear commercially until 1995, by other companies.

 Flyball governor   **1788**

British engineer and inventor James Watt created his first Governor in 1788 after his business associate Matthew Boulton suggested he should pursue the idea and develop it. The Flyball Governor (as Watt named it) was a tapering pendulum based on a version he had developed for his steam engine designs. As the governor's rotation accelerated, fins on top hinges opened out due to the increased centrifugal force being applied and as the fins rose, they were connected to a rotating valve on the central shaft. The valve opened more as speed increased and closed as speed decreased.

 Forstner bit – flat bit   **1874**

Flat wood bits or the Forstner bit as it was originally known, bore precision flat bottomed holes in wood, regardless of orientation. This breakthrough woodworking tool was invented and patented by Benjamin Forstner.

 Fuel dispenser   **1885**

A fuel dispenser or gasoline pump as they are now known, is used to send fuel from a storage tank into containers or vehicles. Sylvanus F Bowser of Fort Wayne, Indiana created the first pump of its type and the term 'bowser' is now synonymous with a device that stores and discharges fluids.

Gas turbine   1791

John Barber was an inventor who is recognised as having applied for and been awarded the first patent for a gas turbine design. Barbers creation used a separate reciprocating compressor that forced air through a gas-fired combustion chamber. The hot jet of gas and air mixture was then aimed through nozzles onto a vaned wheel (much like a water wheel). The output of which was sufficient to power both the compressor and its external load. Future generations of jet engineers based their own developments on Barbers ground breaking invention.

 ## Hydraulic crane   1847

William George Armstrong was an engineer and industrialist who invented the high-pressure hydraulic crane. Armstrong also invented a hydraulic accumulator comprising of a large cylinder filled with water and a piston designed to raise the water pressure in the cylinder, then high pressure water flowed down the connecting pipes to the hydraulic pistons on the cranes and other hydraulic machinery. Armstrong's hydraulic accumulator supplied water with high pressure at up to six hundred pounds per square inch. A few years later following a string of successful developments, he was elected a fellow of the Royal Society.

 ## Hydraulic brake   1918

Invented by Malcolm Loughead, the hydraulic brake is a braking mechanism that uses fluid in pipes (generally ethaline glycol), rather than mechanical cables. The pipes to each respective brake are then terminated into a manifold where one fluid pipe would effectively link a hydraulic brake pedal. When depressed by the driver's foot, the one action will then exert the same pressure across all 2 to 4 brakes on the wheel hubs. This is a far more effective way to control forced pressure in mechanics and is still very much in use today.

Engineering

 ### Industrial robot   1956

An industrial robot is a machine or computer controlled re-programmable manipulator device capable of moving in three or more axes. The co-inventors of the industrial robot were George Devol and Joseph F Engleberger in 1956.

 ### Internal combustion engine   1838

William Barnett was an inventor who applied for a patent for the world's first internal combustion engine design and working prototype. The documents reveal that his submission centered on a double-acting gas engine and detailed a single cylinder, placed vertically with explosions taking place either side of the piston. Barnett's flame method of ignition was very efficient and it was used extensively up to about 1892.

Many inventors attempted to improve on Barrett's design, but none of them overcame the main problem of producing an engine that would be commercially successful.

 ### Jack hammer   1849

Known commonly now as a pneumatic hammer, the jack hammer is a portable percussion drill powered by compressed air. Jonathan J Couch invented the first known drill of its type where the drill bit returns to its original position by a spring.

 ### Jet engine   1930

Sir Frank Whittle was an aeronautical engineer and RAF pilot who filed the world's first patent for a jet engine in 1930 and it was granted in 1932. Whittle was a 24-year-old RAF fighter pilot who first patented a revolutionary type of aircraft engine: the turbojet. It was not built until 1937 due in part to lack of government foresight or backing because its design was so radical a departure from the current aircraft designs. Consequently, he couldn't find enough private backing for the project until 1937, but although he was running his first engine that year, it didn't have its first test flight until 1941. It was a 17-minute debut flight at RAF Cranwell in Lincolnshire in the Gloster Pioneer or G4.

Lift or elevator   1823

British architects Thomas Horner and Decimus Burton designed and installed the world's first public 'lift' or 'elevator' called 'the ascending room' at the Regent's Park Coliseum London, for fee-paying visitors to get a 37-mile panoramic view over the city.

 ### Locking pliers   1924

Locking pliers can generally be fastened into position by the use of an over centre method of lock. On one side of the plier's jaw is a bolt to adjust the width of the jaws and on the other side of the handle is a lever that pushes both sides of the handles apart to unlock them. The inventor was William Petersen of De Witt, Nebraska, later called the 'vise-grip'.

 ### Lock - tumbler   1778

British locksmith, Robert Barron invented the world's first tumbler lock. His patent was submitted in October 1778 and was a critical event in the world of lock making. It was the first lock to feature the 'double acting' principle featuring three tumblers. Prior to Barron's invention, the 'fixed ward' type of lock was the cornerstone of security. A simple bolt was held in a locked or unlocked position by a simple single tumbler catch that located onto the notches at the top edge of the bolt latch. This was a very easy lock to pick and open without a key. Therefore, Baron's three tumbler double acting lock was to lead the way and reset the bar in the world of lock making and high security for years to come.

Engineering

 ### Lock - tumbler - high security   1784

While Robert Barron was fine-tuning his new three-tumbler double acting lock, Joseph Bramah was pursuing a completely different direction in the development of high security locks. In his 'Safety lock' that he patented in 1784, it contained just a small steel tube, with a series of V shaped slots at one end. When the cylindrical key is inserted into the lock, slots accept depressed slides pushed in by the key. The slides must all be precisely the same length otherwise the bolt cannot function. Bramah was so sure that his lock could not be picked that he would give a reward of £200 for anybody who thought they had the ingenuity and knowhow to breach it. His faith in the design was completely justified. It took more than 50 years before a locksmith in the USA claimed the reward

 ### Metal detector   1881

Using magnetic induction to locate metal objects, metal detectors are still used the world over by professionals and amateur metal hunters. British-American inventor Alexander Graham Bell invented the first metal detector whilst living in America and successfully located a lodged bullet in President James Garfield as he lay dying from a fatal gunshot wound.

 ### Micrometer   1638

Sir William Gascoigne was an astrologer and scientific instrument maker from Middleton. He invented the micrometer in 1638, designed to quantify very small degrees of measurement, essential in the world of science and precision engineering. Micrometers based on Gascoigne's design are still made and sold to this day.

 ### Multiple-wheel steam turbine   1884

British scientist Sir Charles Algernon Parsons invented the world's first multiple wheel steam turbine. Parsons first model was attached to a dynamo, generating seven and a half kilowatts of power. His invention allowed the production of cheap and abundant electricity that transformed naval warfare and marine transport.

Phillips head screw   **1936**

Invented by Henry F Phillips, the design of the Phillips head screw lay in its ability to self-centre the screwdriver head and thus it was perfect for using in automated production and manual screwdrivers.

 Radial arm saw   **1922**

Invented by Raymond De Walt, the radial arm saw consists of a circular saw, fixed to a sliding horizontal arm allowing the operator to make length cuts as well as cross cuts. More recent radial arm saws can be turned up to 90 degrees enabling angle cuts to be made easily. A patent was granted in 1925 to De Walt, from Bridgeton, New Jersey.

 Revolving door   **1888**

The principle of a revolving door is that it is essentially a set of 3 to 5 doors, whose ends are fixed to a central vertical axle and then spaced equally apart. This was particularly useful in high rise buildings where standard doors are difficult to open due to differing air pressure, inside and out. Theophilus Van Kannel of Philadelphia addressed this issue with his invention and was granted a patent in 1888.

 Rubber band   **1845**

British inventor Stephen Perry patented the rubber band in March 1845. His early products were made of vulcanized rubber. This ubiquitous stationery item is now used the world over, with billions produced and sold each day.

 Safety lamp   **1815**

Sir Humphrey Davy was a scientist and prominent inventor. The first prototype of Davy's safety lamp was developed in 1815, devised to be lit securely and safely for miners to use. Its design prevented the heat from the flame igniting the methane gas found in pits, as the miners bored deeper into the earth.

Screw propeller   1836

Sir Francis P Smith developed the world's first ship's screw propeller. Sir Francis Pettit Smith was an inventor who with John Ericsson designed and developed the screw propeller. Smith was also instrumental in developing the world's first ship, using a screw propeller – the SS Archimedes. He was later key in influencing Isambard Kingdom Brunel to alter his design of the SS Great Britain, from paddle to screw propulsion after lending Brunel the SS Archimedes for a number of weeks for testing and proving.

Screw threads - standardization   1841

Joseph Whitworth developed and specified the first British national thread standard in 1841. His new standard specified a 55 thread angle, radius of 0.0137329p and thread depth of 0.640327p, p being the pitch. According to Whitworth's chart, the pitch diameter increased in steps. BSW (British Standard Whitworth) is one of three screw thread standards based on the imperial system of measurement, using the same hexagonal nut size and bolt heads, the other two being BSF (British Standard Fine Thread) and BSC (British Standard Cycle). Collectively they are all known as Whitworth threads. The BSW was also adopted in the USA but was replaced when iron bolts used on railways were replaced with steel. But the USA still used BSW right up to the 1970's for some aluminum parts.

Seat belt   circa 19th Century

Sir George Cayley invented the world's first operational seat belt in the early 19th Century. Although Cayley was better known for his huge contribution to the world of aeronautics, he was also responsible for the invention and development of many other significant inventions such as: self-righting lifeboats, caterpillar tracks, prosthetics and optics to name but a few. But he is recognised as being the original inventor who commercialized and applied the use of seat belts, primarily to secure his brave test pilots flying his manned aircraft in the early part of the 19th Century.

Skyscraper   1884
Following the great fire of Chicago in 1871, the city became a launchpad for some groundbreaking architectural experiments and one such design was the world's very first skyscraper. Massachusetts born architectural engineer, William Le Baron Jenney designed what was the Home Insurance Company building, using a steel frame construction rather than traditional timber framed design. This pioneering construction technique enabled Jenney to build much taller buildings, effectively reaching out to the sky, hence the catchphrase and endearing term, skyscraper.

 ## Smoke detector   1890
All smoke detectors work by detecting smoke using either an optical or physical detector, some use a hybrid of both methods, then producing an alarm or signal to notify the presence of smoke. Though, the first practical automatic smoke detector was co-invented by Fernando J Dribble and Francis Robbins Upton using an early physical detection system.

 ## Spark plug - world's first   1888
British physicist Sir Oliver Lodge was a former principal of the University of Birmingham. After many years of research on electromagnetism, Lodge submitted a paper to the British Science Association in Bath, detailing the phenomenon of electromagnetic sparks in varying gap sizes of two wires when supplied with a current. His early discovery developed into electric spark ignition for the internal combustion engine. His two sons later continued his work and later formed the world's first dedicated spark plug manufacturing company 'The Lodge Plug Company'.

Engineering

 ## Steam engine - first steam engine   1698

Thomas Savery was a military engineer from Devon with a great love of anything mechanical and mathematical. Savery patented an early steam engine described on the submission as 'a means of raising water and occasioning motion to all sorts of mill work by the impellent force of fire…'. Savery subsequently demonstrated his invention to the Royal Society in 1699. His invention had no moving parts such as pistons, just taps. Its operation was very simplistic and unreliable by today's standards, relying on the expansion of the steam to push the water up through pipes.

 ## Steam engine - atmospheric   1705

British engineer and inventor Thomas Newcomen constructed the first practical machine to employ the power of steam to generate mechanical output. Newcomen engines as they were known, were eventually sold and used throughout Europe primarily to pump water from mines.

 ## Steam engine - modern condensing   1782

Engineer and inventor James Watt designed and patented the world's first condensing steam engine that produced an uninterrupted rotating motion. Watt's invention, for the first time enabled industrial machines to be powered, being able to be placed almost anywhere so long as a good supply of coal, wood and water was available.

 ## Steam turbine   1884

Sir Charles Parsons developed the world's first steam turbine engine. Following the invention and development of the electric generator by Faraday, the next development was to create a machine that could drive it to generate electricity on a commercial scale. Piston engines were noisy, vibrated a lot and were not reliable, so the steam turbine appeared to be the solution. Today, 75% of the world's power stations still generate electricity using steam turbines based on Parson's original invention.

Tattoo machine 1876

A tattoo machine is a manual hand held device commonly used in the creation of tattoos, providing an indelible marking on the skin. Originally called stencil pens, the inventor of this device was Thomas Elva Edison and it was patented in 1876. This machine was further developed by other inventors years later to include tube and needle injecting of ink, providing a more permanent design on the skin.

 ## Telescopic sight 1630

William Gascoigne was an astronomer who was recreating an optical set-up, when a spider's thread became caught between the two lenses. It was positioned exactly between the two lenses by chance, enabling him to see the web thread and the subject both crystal clear. Gascoigne realized that the precision of the telescope was vastly improved if the web was used as a guide. He further refined his discovery by adding another fine thread crossing at right angles enabling him to centre on an object; thus, the telescopic sight was created.

 ## Tension spoked wheel 1808

George Cayley, a prolific aeronautical engineer of his time was looking for a lighter, stronger type of wheel for his new glider to replace the existing solid wood wheels. His improvement of a new design moved the force balance on his wheels from compression to tension by introducing a series of spoke's spanning from a hub to the rim. This was truly a revolutionary invention that fulfilled all of Cayley's wheel requirements, but the idea didn't really develop further until the advent of the bicycle and powered cars. The rest as they say is history.

Engineering

 ### Thermosiphon - conductive heating system   1828

Thomas Fowler was an inventor from Devon who developed a device he named the 'Thermosiphon' that still forms the foundation of modern central heating systems. The device was the first conductive heating system, but due to severe patent flaws in the way it protected slight changes outside the main design, many other companies soon copied Fowler's brilliant design.

 ### Thermostat   1830

Andrew Ure, a chemist and medical doctor was passionate about the new steam powered machines and factory systems. This love of technology led him to become one of the most prolific inventors of his time. In the huge textile mills of industrial Britain, a constant temperature was required to produce consistent fabric and other goods. This was a difficult task to achieve and therefore the quality of goods was inconsistent. To combat this issue, Ure patented the thermostat that automatically controlled the temperature to improve the consistent quality and standardization of textiles into the future.

 ### Torque wrench   1918

A tool used to apply a precise and specific torque on to a fastener, nut or bolt, it is usually in the form of a standard socket wrench with an additional torque sensing internal clutch that slips as soon as the desired torque or tightness is reached. This pioneering invention, still very much in use today, was invented by Conrad Charles Bahr. He received a patent on his invention much later on March 16th, 1937.

 ### Twist drill bit   1861

A twist drill is a bit with two spiral cut grooves either side of a round steel bar. The spiral grooves produce a helical swarf to aid drilling into wood and steel. Stephen A Morse, the inventor of the twist drill bit, was granted a patent two years after his invention on April 7th, 1863.

Tyre - pneumatic 1845

Briton, Robert W Thompson was only 23 years old when he filed a patent for the world's first pneumatic tyre. He received grants for them in France (1846) and the USA (1847). The design consisted of a low-profile tube of inflated vulcanized Indian rubber between the rim and tyre, providing an 'air cushion' between the vehicle and the ground. Thompson fitted his new invention to several horse drawn coach wheel sets and demonstrated his 'Aerial Wheels' in March 1847 in Regent's Park, London.

 Universal joint - commercial design 1676

Robert Hooke was a prolific 17th century British inventor. Although not a completely new invention in principle, older designs were not practical or efficient in use and so Hooke set about to develop a truly efficient working design. Hooke's Joint is a coupling or linkage that permits the shaft to bend in either direction, commonly used in drive rods that provide rotary motion and transmission. The device consists of two hinge-like joints in close proximity of each other, set at 90 to each other and linked up with a cross shaft.

 Vantablack® - world's blackest material 2014

The darkest and blackest material in the world, was invented and developed by a team of British scientists from Nanosystems®, based in Surrey. Made using carbon nanotubes, four times thinner than a human hair, the material is so dark that it is apparently like looking into a black hole. The material absorbs all but 0.035% of light and can be used for a whole variety of applications such as stealth craft, telescope lining and even combat stealth suits for future 'invisible armies.

Engineering

 ## Vulcanized rubber   1839
Charles Goodyear had a breakthrough when he mixed liquid latex with sulfur when heating it by solar gain or on a stove top. It produces a leather like material that was stretchable and resulted in the first galvanized rubber product. Previously, rubber products were fine until they were exposed to the heat of the summer or a hot surface, where it invariably turned to a gooey mess. Goodyear was subsequently granted a patent for his invention on June 15th, 1844.

 ## WD-40   1953
A water dispersing spray, WD-40 is used both commercially and in domestic homes for loosening and lubricating joints, hinges and for preventing water ingress. WD-40 was invented by Rocket Chemical Company engineer Norm Larsen and two other unknown employees in San Diego, California.

 ## Wellington boots   Early 19th Century
The World's first rubber boots were made by the North British Rubber Company Ltd, taken from a type of riding boot, Wellesley, named after the 1st Duke of Wellington. They were crafted by hessian backed rubber or leather. The Duke was seen many times in portraits of him proudly wearing these boots, allegedly of his own design. More recently, PVC has been used primarily for its cost, durability and effectiveness.

 ## Workbench - portable   1968
Mate-Tool, the world's first truly portable workbench was made by ex-Lotus British car designer Ron Hickman and later licensed to and made by Black & Decker in 1972 as the now ubiquitous 'Workmate'.

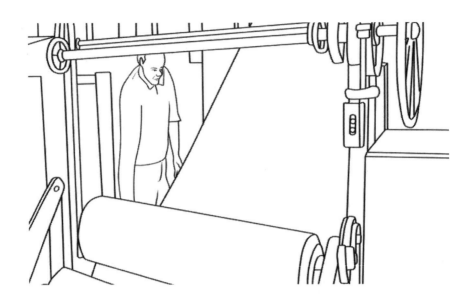

Fabric manufacturing and process

 Acrylic fiber　　　　　　　　　　　　　　　　　**1941**

The Dupont corporation is credited and was awarded patents for the invention of Acrylic fiber in 1941. The fibers are produced from a polymer called polyacrylonitrile. To be classified as acrylic, the polymer must comprise a minimum 85% of acrylonitrile monomer. Dupont patented and trademarked the name 'Orlon' in 1941.

 Elastic fabric　　　　　　　　　　　　　　　　　**1830**

British inventor Thomas Hancock developed fabric elastic. Hancock reformed naturally stretchy rubber in a machine that mixed leftovers of rubber, crafting it in such a way through high-pressure rollers and forming sheets that increased its elasticity. He later developed this process for mass commercial use.

Fabric manufacturing and process

 ### Fibre - man made   1883

British inventor Joseph Swann produced the world's first synthetic or man-made fibre in 1813. Swann extruded a mixture of cellulose nitrate through a grid of small holes. Rayon fibers spun from what is now known as Viscose later replaced these early man-made fibers. Today, the majority of fabrics have a percentage of man-made fibers woven into them to strengthen against wear.

 ### Gore-Tex   1976

A waterproof, breathable fabric, Gore-Tex. A trademarked name is created using an emulsion polymerization process combined with flourosurfactant perflouro-octanoic acid. Gore-Tex was co-invented by Wilbert L Gore, Robert W Gore and Rowena Taylor in 1976 and their invention was granted a patent on March 19th, 1980 then described as a 'waterproof laminate'.

 ### Kevlar   1965

A registered trademark for a light strong synthetic fiber, Kevlaris spun into fabric sheets or ropes for extra strength and durability. Kevlar has a myriad of applications including reinforcement of bicycle tires, yacht sails, body armor and as a replacement for steel in racing car tires. Its inventor is Du-Pont engineer Stephanie Kwolek.

Knitting machine   1589

Reverend William Lee was a clergyman and inventor who produced the world's first knitting machine, used in its original design for centuries after its creation; its key functional concept remains the same today. Lee was said to have created the knitting machine after noting that the woman he was courting at the time, devoted more time to knitting than him. The first version of his machine in 1589 produced stockings using coarse wool, but when he applied for a patent Queen Elizabeth I refused it. So, he developed a second, more refined knitting machine that produced a much finer silkier texture, but his patent application was again denied by the Queen, said to be due to her concern for the job security of thousands of hand knitters throughout her Kingdom.

Disheartened, Lee travelled to Rouen in France with his brother and with the support of Henry IV of France, he commenced stocking production and rapidly grew his business with huge commercial success up to Henry's assassination in 1610. Following Lee's death, a year later his brother returned to Britain and gradually established his fledgling semi-automated knitting industry, often fighting fierce opposition from the hand knitting community.

 Loom - powered   **1785**

British inventor and Industrialist Edmund Cartwright invented the first Power loom that combined threads to make cloth. It was driven by steam, in a factory that Cartwright had set up in Doncaster for cloth manufacture. Being the first, it could always be improved and over the next few years, Cartwright refined his process, benefitting from higher production output and lower maintenance downtime.

 Mackintosh - waterproof coat   **1823**

Charles Mackintosh was a chemist and inventor of waterproof material that led to the creation of the Mackintosh raincoat. It was through experimentation with the by-products of tar that led Mackintosh to discover a waterproof covering for fabrics and eventual patent application for his process. Essentially, Mackintosh's invention was based on gluing two fabrics together with Indian rubber, made soluble by Naphtha. For many years to come, the term 'Mackintosh' or 'Mac' became the household name for a raincoat.

 Mercerized cotton   **1850**

John Mercer was an inventor from Great Harwood. Mercer's treatment of cotton fabric using sodium hydroxide to swell the yarn and give it a lustrous sheen is now known as 'Mercerizing'. Mercer's process is still used internationally in the production of fabric.

Pinking shears — 1893

The first patent associated with pinking shears was granted to Louise Austin of Whatcomb, Washington on January 1993. She devised this variant of the common scissor to improve on pinking irons, previously the only way to pink or scallop fabric materials.

Polyester — 1941

John Whinfield and James Dickson were scientists, who together invented polyester in 1941 and because of its flexibility in making artificial fabrics it became an overnight success. Polyester is made from oil, or to be more precise, polyethylene terephthalate that originates from petroleum. Its invention was declared a wartime secret due to its many uses during severe raw material shortages in WWII.

Rayon - viscose — 1892

Of all man-made fibers in the world today, Rayon is the oldest of them all. It was the first viable manufactured fibre. English Chemist Charles F. Cross, and his collaborators Edward John Bevan and Clayton Beadle, discovered the process of making Viscose in 1892.

Rayon - acetate — 1895

The cellulose acetate manufacturing process was invented and produced initially by Charles Frederick Cross and Edward John Bevan (of Viscose fame). It was used by the pair as an alternative to their original invention of Viscose Rayon, due to its differing texture.

Sewing machine — 1790

Thomas Saint, the inventor, is generally recognised for developing and making the world's first sewing machine design, but like so many inventors, did not adequately market or sell his invention. His machine was designed to be used on canvas or leather material. The device itself featured a feed mechanism, overhanging arm, a looper and vertical needle bar, using the chain stitch method where the machine used a single thread to make simple stitches on the leather or canvas. Saint's Sewing machine was very advanced for its time, but its undeveloped and untested concept meant that it would need much refinement and development over the ensuing years to become a truly practical proposition.

 ## Spandex   1959

A synthetic fiber known for its remarkable elasticity, spandex is generally woven as sports clothing for exercising or gymnastics. Spandex was invented by a Chemist Joseph Shivers, working for Dupont in 1959.

 ## Spinning Jenny   1764

inventor, carpenter and weaver James Hargreaves, developed an improvement on the established spinning wheel by creating the 'Spinning Jenny'. Hargreaves' device was a hand turned spinning machine that powered eight spindles to create an eight-threaded weave. Advanced models later in its development used up to one hundred and twenty spindles. Like so many technological advancements, Arkwright's machines were a threat to traditional fabric workers, so much so that in 1768 a group of spinners destroyed Hargreaves' Spinning Jenny machines after breaking into his house, fearing their jobs would be lost forever.

 ## Spinning frame   1768

The Spinning frame was invented and developed by industrialist Richard Arkwright and it became the keystone of the global textile boom at the start of the industrial revolution. Arkwright's spinning frame spun thread that had a much tighter weave than previous incarnations and because of the new fabric density, was far stronger. Britain became known as the 'Workshop of the world' through this single key invention, transforming the textile industry in England's North West.

Fabric manufacturing and process

 ## Spinning mule   1779

Industrialist and inventor Samuel Crompton created the spinning mule in 1779. Crompton's invention combined the moving frame of the spinning Jenny with the water frames rollers. His device gave the operator greater control over the weaving process, enabling many different sizes and texture of yarns to be used.

 ## Transverse shuttle   1846

A method of driving a bobbin on a sewing machine to create lockstitch, the transverse shuttle is one of the earliest known examples of bobbin drivers. The shuttle supports the bobbin in a ship shaped shuttle and counters the returning shuttle along a level horizontal shaft. It was subsequently patented by Elias Howe.

Food and drink

 Bread slicer   **1927**

Frederick Rohwedder invented the commercial automatic bread slicer in 1927. The device sliced a bread loaf into a pre-sliced package, then automatically wrapped the bread for customer convenience.

 Breakfast cereal   **1863**

Granula was the first breakfast cereal, invented by James Caleb Jackson, founder and owner of the Jackson Sanitorium in Dansville, New York. Unfortunately, his Granula cereal was not successful due to the impracticality of having to soak the large bran chunks overnight to become tender and ready to eat.

Food and drink

 ### Champagne or sparkling wine   1662
Christopher Merrett was a physician and scientist from Winchcombe in Gloucestershire, England. He was the first person to document the intentional addition of sugar for the manufacture of sparkling wine. In his eight-page paper, Merrett refers to a secondary fermentation process, a chemical reaction occurring when the bottled alcohol underwent increases in temperature, thus producing carbon dioxide. This process now forms the basis for all Champagne production. The French have always maintained that Sparkling Wine or Champagne (as it is now known), was invented by the Monk Dom Perignon in 1697 by accident. But Interestingly, the earliest French documentation is 1729, an incredible 67 years after Merrett's invention.

 ### Champagne bottle - design origin   1652
Admiral Robert Mansell discovered that adding iron and magnesium to the glass making process, made a stronger glass to contain high gas pressure. Sparkling wine inventor and Briton, Christopher Merrett, needed a bottle that would not explode under pressure from his new sparkling wine and later adopted Mansell's design. Contrary to common belief this was not a French design and was created long before French sparkling wine started being produced. Mansell's original bottle design also predated Merrett's sparkling wine.

 ### Cheese burger   1924
A hamburger with a cheese slice added to the top of a hot beef patty, melting the cheese on top, the cheese burger is renowned the world over as a staple fast food choice. Its inventor is believed to be Lionel C Sternberger from Los Angeles, who invented it in 1924.

 ### Chocolate bar   1847
Joseph Fry made the world's first chocolate bar, developed and created by his company JS Fry & Sons of Bristol in 1847. Up until this point, chocolate had only been available as a drink. The secret was to mix cocoa powder

with cocoa butter and sugar, ensuring the product remained solid at room temperature, but melting in the mouth when eaten. It was billed originally as 'chocolate delicieux a manger', or delicious to eat.

 ## Coffee houses - coffee shops   1650

The real coffee shop revolution didn't start in recent times with the large multi nationals. In fact, it first started in 1650 in Oxford and following its success moved to London in the Cornhill district, taking this new drink to the gentry of this bustling city. It also became the birthplace of the London Stock Exchange. Prior to coffee shops, traders had carried out their business in taverns that were by their nature, raucous and unpleasant. By comparison, Coffee prevented drowsiness and sharpened the mind, spawning the beginnings of what is now known as the stock market.

 ## Condensed milk   1856

The process required to produce condensed milk was invented by Gail Borden and was used by soldiers during the American civil war. It is essentially cow's milk with the water content removed and sugar added, producing a sweet, rich, creamy substance than is capable of lasting for many years once opened.

 ## Cornish pasty   1509

With a history dating back to the original tin industry of Cornwall, the Cornish pasty was developed and made by wives to feed their hungry tin mining husbands and sons. Affectionately known as an 'oggie' by locals, the traditional Cornish pasty consists of beef skirt, potato, onion and turnip, wrapped in a case of pastry and deep crimped around the edges. The crimped crust served as a handle to hold the hot pasty when down in the mine and the thick short crust pastry acted as a superb insulator, keeping the hand-held meal hot before eating. After the demise of the Tin Mining industry, many experienced miners emigrated and took their recipes with them. You can now get a great pasty in Argentina, Australia, New Zealand, Brazil and many other South American countries.

Food and drink

 Cotton candy   1897

Co-invented by candy makers John C Wharton and William Morrison, from Nashville Tennessee, cotton candy (or candy floss as it is known in Britain) is a soft fluffy confection, that is created by heating and spinning sugar into very thin threads to create the illusion of a giant cotton ball.

 Diner   1872

A diner is universally regarded as an American style café, categorized by its wide range of food, old style service and nostalgic atmosphere. Although, the forerunner to the ubiquitous fast food restaurants started when Walter Scott, a Providence Journalist, got serious about selling food and drink on the streets. Armed with a horse drawn delivery cart, he took food to the people using his Diner. It was never patented but was developed substantially over the years into what we now know as an all-American diner.

 Fish and chips   1860

For many years, Fish & Chips have been the national dish of Great Britain, often symbolizing its great cultural and culinary heritage and like the Union Flag, recognised as British the world over. Fish and Chips were first served together as a single dish around 1860 with both Manchester and London laying claim to being first. But fried fish and fried potatoes had been in existence in Britain for hundreds of years prior to the joining together of this great gastronomic marriage.

 Fizzy drinks - carbonated drinks   1772

Eighteenth century Scientist and clergyman Joseph Priestly, was the inventor of carbonated water. He did this by suspending a tub of water above a large fermenting ale container at a local brewery in Leeds and the water above rapidly developed a pleasurable gaseous syrupy and acidic taste. He published a description of how to produce his soda water in 1772. Several years later, Johann Schweppe created Schweppes and started manufacturing fizzy drinks using the Priestly method.

Frozen food 1929

Invented and developed by Clarence Birdseye, frozen food was and still is simply any foodstuff preserved by the method of freezing. By effectively turning water to ice in food, the process prevents most bacterial growth to occur, slowing down possible decay and chemical reactions to enable food to be prepared and then stored for weeks or months until ready to use.

 ### Gin and tonic circa. 1820

Widely regarded as the world's most civilized cocktail, G&T was created by British soldiers based in India. Tonic water contained Quinine a malaria preventative that had a bitter after taste and the gin was used to sweeten the drink.

 ### Hamburger 1916

Despite its name, the hamburger generally has no ham in its ingredients. Instead it is normally a beef meat and oat flat disc shaped patty, that is typically grilled and is placed between two halves of a toasted bread roll. Its inventor was Walter Anderson who co-founded the famous American fast-food chain, White Castle in 1921.

 ### Ice cream maker – hand cranked 1843

Used to produce small quantities of ice cream at home, the ice cream maker stirred the mixture using a hand crank and more latterly an electric motor. The operator then chilled the ice cream by freezing the mixture, whilst stirring occasionally, to prevent the formation of ice crystals. It was Nancy Johnson from New England who invented and patented the first-hand cranked ice cream maker. Though, lacking the resource to develop and market it herself, she sold the rights to a kitchen wholesaler who had a huge success with Johnson's original design.

Food and drink

 ## Potato chips　　　　　　　　　　　　　　　　　　　1853

Thin slices of potato that are baked or deep fried until crispy, potato chips, or crisps as they are known in Britain, served as an appetizer or side snack, often with added salt for flavor. The original invention and recipe was devised by George Crum from Saratoga Springs, New York and the fried potato slices were originally known as Saratoga chips.

 ## Sandwich　　　　　　　　　　　　　　　　　　　　1762

John Montague, the fourth Earl of Sandwich, coined the word 'sandwich' according to historic legend. It is doubtful that he actually invented what we now know as a sandwich, but he certainly made it popular and gave rise to its name. It was apparently in 1762 when he was said to have asked for meat to be served between two slices of bread, in order to avoid interrupting a gambling game. This story may have been put about to create adverse publicity by his rivals, but the term sandwich stuck and is now a household name used worldwide. The Earl had no actual connection with the English coastal town of Sandwich, only by title. Proving that British hereditary titles can be very confusing. The Sandwich Isles were also named after the fourth Earl of Sandwich.

 ## Soft whipped ice cream　　　　　　　　　　　　　1949

Prior to a career in politics, ex-Prime Minister Margaret Thatcher took on a junior research role as part of a development team at food manufacturer J. Lyons and Co, putting her Oxford chemistry degree to good use. The main focus of her role was to develop a method to introduce more air into ice cream to produce a smoother whipped texture that could be forced through a tube, forming part of an ice cream delivery system. This system proved popular and was the catalyst for '99 ice creams' and brands such as 'Mr Whippy".

Soft served ice cream   1938

Soft serve ice cream is a unique frozen dessert similar to normal ice cream, but softer in texture. The co-inventors credited with this easier to scoop ice cream, were J F McCullough and his son Alex. This was to be the catalyst for further development that would eventually lead to machine fed soft whipped ice cream.

 ### Tea bag   1903

Invented and developed by New York coffee and tea merchant, Thomas Sullivan, the tea bag is a small absorbent silk or paper sealed pouch containing tea leaves for brewing tea. The idea came about after he discovered that small pouched samples of tea he sent to customers, were being used intact to brew in pots.

 ### Tin can   1810

Up to 1810, food could only be preserved by either salting heavily or storing for limited time in glass jars, by cooking it for hours first to help sterilize it. However, British Merchant Peter Durand took the jar preservation method a step further and decided to use a tin can. The main problem was that either a sharp thick blade or a hammer and chisel were required to open the cans. It would be another 48 years before American Ezra Warner invented the first tin can opener.

Household and general Innovations

 Air-bubble - packing 1957

Known more commonly as 'bubble-wrap', air bubble packaging is a transparent, flexible air-bubble protection and packaging material. Marc Chavannes and Alfred Fielding were the co-inventors of air-bubble packaging.

 Baby buggies - strollers 1965

Aeronautical designer and test pilot Owen MacLaren, produced the world's first lightweight aluminum pushchair that folded using one hand, developed from his expertise in designing the folding undercarriage for wartime Spitfires. Following testing, he then started manufacturing them in 1967 and now the name MacLaren is synonymous with the folding baby buggy.

Backpack - internal frame   **1967**

Comprising a frame constructed from tubes or strips of metal or plastic that molds to the users back, the backpack provides a canvas or other waterproof outer skin and is a convenient means to carry often heavy and cumbersome loads such as tents, sleeping bags and supporting items. The backpack was invented by Greg Lowe, the founder of Lowepro in 1967.

 Ball barrow   **1974**

In 1974, James Dyson, the prolific British inventor, designed the ball barrow. It featured a steel frame supporting a plastic molded hopper and a front mounted spherical plastic wheel. It was developed through the restrictions of conventional wheelbarrows using thinner pneumatic front wheels, that were less controllable and prone to getting punctures. The ball wheel also gave the barrow stability in soft ground surfaces making it more stable laterally under load on more uneven ground. The Ball Barrow was awarded the British Design Innovation Award in 1977.

 Ballpoint pen   **1888**

A writing instrument with an internal vessel of ink and a round point, the ball point pen dispensed its ink through its tip from a rolling action across the paper. The inventor of this device was American Tanner, John Loud from Weymouth, Massachusetts. Loud took out a British patent in 1888, although it wasn't until 1935 when Hungarian, Laszio Biro improved on Loud's design that left the page smudge free after use.

Household and general innovations

 Banknotes - first public issue   **1797**

The Bank of England issued the world's first commercial banknotes in 1797 after the small-scale Napoleonic invasion of Fishguard in Wales, caused jitters amongst its hierarchy and forced the bank to issue IOU bank notes. The plan was for Napoleons second rate army to land near Bristol, destroy it and then move on into Wales, marching North towards Liverpool and Manchester. It is recorded that the Bank of England took fright at the Fishguard Invasion because although unsuccessful, they feared a similar more successful invasion might happen again. The bank then decided not to use the existing gold and silver coinage anymore and instead printed banknotes on paper in one- and two-pound denominations.

The banknotes were effectively IOU's with the declaration 'I promise to pay the bearer on demand the sum of…' still printed on British banknotes. The paper notes were printed on special watermarked paper to prevent copying. From this day on, the world's first bank notes appeared in general circulation.

 Bathtub - cast iron enamel   **1870**

Enamel cast iron bathtubs were developed and invented by Briton David Buick. He moved from Great Britain to Detroit Michigan with his parents as a young man and worked in the sanitary and plumbing industry. He developed and invented the internationally successful process for bonding porcelain enamel to cast-iron bathtubs. In 1903, pursuing his interest and love of motorcars, Buick founded the Buick Motor Company. The rest is history.

 Can opener - rotary   **1870**

The can opener is a tool designed to open metal cans. William W Lyman from Meriden, Connecticut, invented a rotary hand cranked version that you can still buy a modern variant in shops today. He was granted a US patent on July 2nd, 1870. His design consisted of a wheel with a serrated edge that pierced and cut the can top as it was cranked around the tin lid.

Coffee percolator 1806

Invented by Benjamin Thompson Rumford in 1806, the coffee percolator is a type of vessel used to brew coffee, using hot water to percolate through the grounds of coffee. One of the key components of Thomson's invention was a metal sieve that strained the grains of coffee and prevented them entering the cup.

 Corkscrew 1795

Samuel Henshall patented the world's first corkscrew. He was an English clergyman and it is thought that he designed the corkscrew through his frustration of releasing corks from church wine bottles, which until this point was very haphazard and often dangerous to the 'de-corker'.

 Correction fluid 1951

Typically, an opaque white liquid that is applied to paper to overlay a written or typed mistake, correction fluid was an extremely useful product when typing. The fluid was designed to dry quickly and to be overwritten or overtyped. Correction fluid was invented by Bette Nesmith Graham in 1951 under the brand name of 'Mistake out'.

 Clothes hanger 1869

A clothes or coat hanger is a product, shaped like human shoulders and created to hang shirts, coats or jackets, thus preventing creases and wrinkles. On most clothes hangers, there is a lower bar to hang skirts or trousers. This universal device was invented by O A North from New Britain, Connecticut.

 Deodorant 1941

Applied to the body to diminish body odor produced by bacterial breakdown originating from perspiration, deodorant was first invented by inventor Jules Montenier to counter this age-old personal hygiene problem.

Household and general innovations

 Diaper 1946

Used as an absorbent garment for infants and adults, the diaper or nappy was invented as a 'dampless' or 'waterproof' diaper' by Marion Donavon. First retailed in 1949 in Saks New York Fifth Avenue store, patents were later granted in 1951 and Donavon sold the rights to the 'waterproof diaper' for $1 million.

 Drinking straw 1888

A drinking straw normally consists of a small tube and is used for the transfer of liquid to the mouth. In 1888, Marvin Stone used a sheet of paper and wrapped it around a pencil. After coating it with wax, he prevented the straw leaking, stopping the paper becoming soaked and inoperative by the liquid.

 Electric blanket 1912

Invented by American physician, Sydney Russell, the electric blanket is a combined blanket and heating element positioned on the top sheet of a bed. Russell's earliest design was an electric under blanket produced in 1912.

 Electric cooking utensil 1874

St George Lane-Fox, a prolific British inventor and electrical engineer took out a patent for an 'electric cooking utensil' in 1874. There is little information on how the patent was commercially exploited, but it is clear that this was the very first example of electric cooking, as we know it today.

 Electric fan 1882

An electric fan employs an array of blades, generally powered by an electric motor to produce forced airflow for cooling on the earliest models, but also for heating in more recent times. The first electric fan was invented by Schuyler Wheeler from New Orleans.

Electric stove (cooker)   **1859**

A large kitchen appliance that converts electricity input into regulated heat is known as an electric stove or cooker. Canadian inventor Thomas Ahearn is often credited with this creation in 1882. Although, a patent was awarded much earlier on September 20th, 1859, to American inventor George B Simpson.

 Eraser   **1770**

Edward Nairne was an optician from Sandwich. In 1770, one of the first references to the word 'rubber' was discussed, when Nairne sold natural rubber cubes at his shop in Cornhill. Nairne's cubes were erasers, sold for the extraordinary sum of £3 per cube. Edward Nairne is widely credited with creating the first rubber eraser.

 Flushometer   **1906**

A water pressure process that employs an inline handle to flush urinals and toilets, a flushometer is generally fed by mains water pressure rather than a connected water tank. This system ensures that there is a quicker recycle time between flushes. The flushometer was invented in 1906 by businessman and inventor, William Elvis Sloan.

 Formica   **1913**

Formica was co-invented by Westinghouse employees, Daniel J O'Connor and Herbert A Faber in 1913. Formica is a hard, resilient laminated plastic used for cupboard doors and work surfaces that require a heat-resistant finish.

Friction match   1827

John Walker was a British chemist who invented the world's first friction match. Like so many great inventions the friction match principle was discovered purely by accident, when a wooden splint encrusted with chemicals burst into flame when scratched across Walker's fireplace hearth at his home. He further developed his discovery and made the first friction match in the world. Walker's friction matches transformed the portability and application of fire. Against the advice of friends, Walker failed to patent his invention and consequently in 1829, Samuel Jones from London copied his method and launched 'Lucifer's', a blatant copy of Walker's 'friction lights'.

Fridge - principle of heat exchange   1748

William Cullen was the British physicist and chemist from Hamilton. Cullen demonstrated the first known artificial refrigeration process at the University of Glasgow in 1748 but did not use his discovery for any commercial or practical purpose. To this day, Cullen's principles of heat exchange, employing cooling by rapid expansion of gases, are the main source of refrigeration in all modern systems.

Jeans   1873

Generally made of denim, jeans became popular in the 1950's, but its roots go back to 1873. Levi Straus and Jacob Davis co-invented the process of using copper rivets at the stress points in workers pants, such as pocket corners and the top of the button fly. Originally, they came in brown or a heavy denim blue fabric and the rest is history. They were granted a patent in 1873.

 Hairspray   1948

A beauty product, hairspray is used to maintain stiffness in hairstyles and is typically sprayed to maintain styling for long periods of time. Hair spray was invented and developed by Chase Products Company in 1948.

Iron – electric   1881

The small household appliance known as an electric iron was invented and patented by Henry W Seely of New York. Designed exclusively to iron out wrinkles and creases and to press clothes.

Ironing board   1858

Designed with a heat resistant top on a folding narrow table, the ironing board was designed to be used with an electric iron to neatly support clothing to be pressed at an adjustable height to suit the user. It was co-patented by its inventors, William Vandenburg and James Harvey of New York.

Kettle - electric   1891

The electric kettle was invented to satisfy the huge demand and need of British tea drinkers during the late 19th Century and this invention was credited to Compton & Co in 1891. Several improvements were made over the years, including the British Birmingham based Swan Company who in the 1920s developed the world's first kettle with a built-in heating element.

Kettle - automatic electric   1955

Founders Bill Russell and Peter Hobbs of Russell Hobbs Co Ltd altered the course of standard electric kettles in the 20th Century by designing and introducing their revolutionary vapour controlled K1 automatic electric kettle design in 1955. This British made and designed K1 was the world's first fully automatic kettle. When it was first launched, demand far outstripped production such was its popularity. Before the K1 was launched, old style electric kettles suffered from boiling dry when unattended, creating fire hazards and were a danger to the home, not to mention the element burning out if left without water cover and needing to be replaced each time this happened. This ground-breaking design was so effective and successful that it is still used the world over in all electric kettles sold today.

Household and general innovations

Lawnmower   1827

British inventor and landscaper Edwin Beard Budding developed the world's first mechanical lawnmower (19 inch) and because of its initial high price to make and buy, it remained the preserve of the very rich until such time that mass production offered lower cost alternatives.

Lawnmower - steam powered   1893

James Sumner, a British inventor from Leyland in Lancashire, patented the world's first lawnmower powered by steam. Known as the Leyland Steam Lawnmower, it was operated by just one skilled Engineer, who was required to be an expert in steam power, its maintenance and operation. The boiler had to be constantly topped up with water and the machine needed regular meticulous maintenance like any steam driven machine, making running costs very high. In 1895 after moderate success in sales, Sumner renamed his company the Leyland Steam Motor Company a predecessor of the infamous British Leyland Motor Company of 1970's fame. It was not until the introduction of widely available petrol-powered internal combustion engines, that powered lawnmowers became a large market commodity.

Luggage - tilt and roll   1988

In 1988, Northwest Airlines pilot, Robert Plath invented tilt and roll luggage after watching countless passengers struggling with their suitcases in 'no-trolley zones'. His invention consisted of a suitcase with two wheels on the bottom back section of the case and at the top back side there was a telescopic handle used to tilt and pull the suitcase.

Masking tape   1925

Used as an easy to tear, pressure-sensitive self-adhesive tape, masking tape was invented by Richard G Drew, an employee from the Minnesota Mining and Manufacturing Company, known generally as 3M. His first prototype consisted of a 2-inch-wide tape with a light pressure sensitive adhesive, that when pulled off the contact area, it did not mark or damage the surface.

 Meccano 1898

Frank Hornby was an inventor who decided to create a system of parts including nuts and bolts, spanners and screwdrivers, to keep his two young sons occupied through the long winter days. Hornby wanted his boys to be able to recreate the huge cranes that enthralled them so much at their home in the port of Liverpool. Hornby set about developing and commercializing his kit of parts and started to market and sell them under the brand name 'Mechanics Made Easy' in 1901. The name 'Meccano' was patented in 1907.   Hornby's first Meccano factory opened its doors in Liverpool the same year, but as demand was so huge, a much larger plant quickly replaced it. The popularity of Meccano soon spread across the globe and at one point in 1951 just one of Meccano's many plants, the Bobigny factory produced more than half a million boxes in a day. Meccano is still sold in huge quantities the world over; such is its continued demand.

 Microwave oven 1945

Cooking with microwaves was discovered by Percy Spencer whilst working on a live radar system. He noticed an abnormally hot feeling throughout his body and spotted that a candy bar in his pocket had started to melt. The candy bar had been heated by microwaves from the RADARs magnetron. Following further investigation, Spencer deliberately cooked popcorn, leading to his first demonstration of the world's first microwave oven in 1947.

 Mousetrap 1899

Born in Leeds, James Henry Atkinson was the inventor who created and patented the spring-operated mousetrap called 'the little nipper'. His design is still sold to this day and many regard the design as being the most effective means of catching mice.

Household and general innovations

 Paper clip   **1867**

Designed to attach sheets of paper together, the paper clip allows the sheets to be easily detached as required. This simple but hugely popular item of stationery was granted a patent to its inventor, Samuel B Fey in 1867.

 Paper shredder   **1909**

Invented by prolific inventor Abbot Augustus Low of Horseshoe, New York, the paper shredder is used to cut paper into cross cut strips or small pieces mainly for privacy and security reasons. Lows paper shredder invention was granted a patent as a 'waste paper receptacle' on August 31, 1909.

 Pencil eraser   **1858**

An article of stationery attached to the opposite end of a pencils tip. The pencil eraser is typically made from gum or synthetic rubber, used to rub-out or erase mistakes by pencil on paper. Inventor Hymen Lipman was granted the first patent for the idea of attaching a rubber on the end of a pencil.

 **Pram - perambulator or baby carriage**   **1733**

William Kent designed the first portable pram or baby stroller. Kent was asked to build a device that would transport his children and he duly designed and built a three-wheeled frame with a basket and handles. The baby or infant sat in the basket. Kent's design is remarkably similar to the three-wheeled prams available today.

 Pressure washer   **1927**

Invented originally by Frank Ofeldt, the pressure washer is a mechanical sprayer that outputs a high-pressure spray. It can be used to clean grease, grime or dust from most surfaces and can even be used to remove loose paint. It was originally called a 'high-pressure jenny'.

Razor - stainless steel long-life blades   1962

Wilkinson Sword made the world's first commercial stainless-steel long-life blades. Early carbon steel blades either had to be sharpened regularly or disposed of after every shave. Wilkinson Sword's blades, although more expensive, could be used for a week before the need to replace them.

 ### Razor - electric   1928

Invented by Colonel Jacob Schick, the electric razor has a vibrating, revolving or oscillating set of blades to remove unwanted bodily hair. One of the key benefits of the electric razor is that it provides a dry shave, without the use of soap, foam or creams. The first electric shaver was powered by a micro AC motor and plugged directly into the mains with no batteries. Subsequent electric razors used rechargeable batteries powering a small DC motor.

 ### Refrigerator - vapor-compression   1805

The process of removing heat from an enclosed space or from a substance is known as the process of refrigeration. American inventor Chris Evans invented the vapor compression refrigeration machine in 1805. His process removed heat from an enclosed area by recycling vaporized refrigerant, where it would travel through a compressor and condenser, eventually reforming as liquid to repeat the process again.

 ### Remote control   1898

An electronic device used to operate a machine from a short distance, the remote control was initially conceived and invented by Nikola Tesla and demonstrated in Madison Square Garden at the Electrical Exhibition. His first remote control operated a model boat powering through the water, controlled by radio signals. Tesla was granted a patent for his invention in 1898.

Household and general innovations

 **Safety pin**   **1849**

British inventor Charles Rowley was regarded as the creator of the safety pin, similar to those in use today. A sliding clasp covered the point of the pin to protect the user with a simple spring joint at the other end to aid opening and closing of the device. Walter Hunt (USA) also made a similar device at around the same time but sold his design to a US company who ended up making millions of dollars in profit over the years to come.

 Safety razor   **1901**

Designed to protect the skin of the user, the safety razor was invented by travelling salesman King Camp Gillette from Wisconsin. Gillette created a disposable safety razor that was attached to a reusable handle. Prior to his invention, blunt razors would be taken to the local barber to sharpen, but Gillette's double-edged disposable blades provided a clean, uniform shave. Gillette was granted a patent in 1904.

 **Sewing machine - lock-stitch**   **1833**

Walter Hunt was the inventor of the lock-stitch process, still seen on modern day sewing machines. Consisting of two cotton threads, the upper thread runs from an upper spool and through the eye of the needle, whilst the lower thread is wound on to a bobbin, located in a case in the lower section of the machine. The first lock-stitch machine was invented by Walter Hunt, although he failed to patent it at the time. Subsequently in 1846, Elias Howe was granted a patent.

 Shaver - electric   **1913**

British inventor G.P. Appleyard patented a 'power driven shaving appliance', although the design was revolutionary, the device could not be manufactured because motors could not be made small enough at the time.

Shoelaces 1790

Briton Harvey Kennedy is credited as being the inventor of the modern shoelace. He also patented the 'shoe fastening lace' in 1790 and records suggest that he profited a great deal from the proceeds.

 Slot machine 1887

Devised as a semi-automatic casino gambling machine, the 'one-armed bandit' was invented by Charles Fey of San Francisco. He created a simple mechanism that was automated with three spinning vertical discs, side by side in a screen, each containing five symbols. The gambler won a set sum, if all three discs matched once spinning stopped. Payout was automatic. To this day, almost every casino in the world still sports variants of the original slot machine.

 Steel pen 1780

It is documented that Samuel Harrison, a manufacturer of split pins, made a steel pen for Dr Joseph Priestly in 1780 and is the first recorded use of steel as an ink pen rather than a feather quill.

 Stove - gas 1826

With the advent of coal gas producing companies being firmly established in most major towns and cities in Britain, James Sharp spotted an opportunity to replace the existing coal and wood fired ovens with something altogether more practical. Having constant fuel on tap, the gas stove was developed and patented by British inventor James Clegg. The gas stove was a more efficient method to cook food and it is testament to James Sharp's original design, that gas stoves are still being used today and are often still the first choice for professional chefs the world over.

Household and general innovations

 Swivel chair   **1976**

A revolving or swivel chair normally uses a single upright base or leg to allow the spindle attached to the central seat to rotate. Some swivel chairs have wheels on its base to enable the user to effortlessly move the chair around. To this day, swivel chairs are ubiquitous across the world in offices and home studies. Its inventor Thomas Jefferson used an English Windsor chair as a prototype . Legend has it that Jefferson used his own swivel chair when he drafted the United States Declaration of Independence.

 Tea maker - automatic   **1902**

Albert E. Richardson was a British clockmaker who invented and patented the world's first automatic tea-maker. Methylated spirits were lit by the automated striking of a match, thus heating the hot water and this action then started the chime of a large portable alarm clock bell when the tea was ready. The Automatic Water Boiler Co Ltd was contracted to make the machine.

 Teddy bear   **1902**

Usually stuffed with a soft fabric cotton filling and clothed with soft fur, the teddy bear became an enduring childhood cuddly toy that has lost no favour amongst children and adults alike. The original inventor of this soft toy legend was Morris Michtom who owned a Brooklyn store in New York. The rumor of how the bears were called 'Teddy' is believed to originate from the president Theodore 'Teddy' Roosevelt, who gave his permission to use his name for the cuddly furry bears after sparing the life of a Louisiana black bear cub while on vacation.

Toaster - electric   **1893**

British inventors, Alan MacMasters and Rookes Evelyn Bell Crompton's main specialties were the fields of Municipal Electricity Supplies and lighting and they founded the company Crompton & Co. British inventor Alan MacMasters approached Crompton with an example of a device for heating bread by the use of an electrical element. Masters and Crompton worked together to perfect the design and it later went into production that year. This was the world's first electric toaster and it was named 'The Eclipse'.

 ### Toaster - pop-up   1919

Following on from the original toaster invented by British inventor Alan MacMasters, American inventor Charles Strife, took this idea a stage further and automated the process by creating a toaster that popped the toasted bread out of the toaster slots after the elements had heated the bread for a variable set time. Thus, reducing the risk of burnt toast.

 ### Toilet - flushing   1596

John Harrington installed the first non-gravity flushing toilet at his own home in Somerset. Harrington made one for Queen Elizabeth I (his God Mother) and one for himself. He only ever produced these two and was often mocked for his invention, so much so that it was not until 1775 that Alexander Cummings made huge progress in the design of his water closet and momentum then gathered.

 ### Toilet - s-trap   1775

British inventor Alexander Cummings filed the world's first patent for an 'S-Trap'. The flush toilet was emerging in its modern form as a device. But it was Cummings' invention of the 'S Trap' that really advanced toilets in the mid 18th Century and is still in use worldwide today. This simple device uses the standing water in the bottom 'U' section of the S trap to act as a seal for the bowl outlet to stop the escape of foul air entering the room from the sewer.

In the mid 19th century, it also prevented vermin and all manner of wildlife from entering the room from the toilet bowl. As an added precaution, Cummings design also incorporated a sliding valve just above the valve in the bowl that remained shut until the toilet was flushed.

Household and general innovations

Toilet paper – mass produced   1857
Used to maintain personal hygiene, toilet paper is a soft paper product. It can also be used for a myriad of other uses where liquid absorption is needed, such as spillages. The first patent applied for a toilet paper solely for use to wipe a person's buttocks was from New York based entrepreneur, Joseph Gayety. His design included aloe-soaked paper tissue dispensed from tissue-like boxes.

Toothbrush   c1770
William Addis designed and made the world's first toothbrush while in Newgate prison after being caught up in riots and then arrested. The businessman used an old bone and the hair of a pig to make the first toothbrush. Necessity is the mother of invention and following release from prison, he built a global powerhouse and the brand still exists today, known as Wisdom Toothbrushes.

Umbrella   1868
Samuel Fox was a British industrialist and is best known for his invention of the Paragon steel umbrella frame. His U section steel frame was far superior to his competitors at the time, who all made their frames from whalebone. His design was so successful that production of Fox frame umbrellas only ceased production in Britain in 2003 and the company ceased trading in 2008.

Vacuum cleaner - manual   1860
Using a partial vacuum to suck up dirt and dust, a vacuum cleaner design was submitted for patent approval in 1860. Inventor Daniel Hess of West Union, Iowa was subsequently granted a patent later that year in July. His machine was designed with a rotating brush and incorporated sophisticated bellows in order to suck the dust from the floor into a separate container.

Vacuum cleaner - powered 1901

Herbert Cecil Booth was observing a railway coach in the process of being cleaned whilst out of service. The cleaner used a machine that just blew the dust away and in the event, ended up making more mess inside the coach than was intended. This prompted him to create a device that would work in reverse and suck up the dust and deposit it into a container, thus leaving the coach dust free. While developing this new machine, he tested his theory by placing dusters or handkerchiefs on the floor and then observed that dust was sucked up and stuck to the cloth on both sides. This was the breakthrough that he was looking for, because it prevented dust flowing through to the motor and causing all manner of problems. Once he fully developed a filtered version, he then set about starting a company using hoses from a purpose converted van on the street and taking the hoses through doors and windows to clean houses. This was the birth of the first Vacuum cleaner, as we know it.

 Vacuum flask 1892

Sir James Dewar, a British Cambridge based professor and prominent figure of the Royal Institution, invented this simple device. He didn't create it to keep his tea or coffee hot, but to act as an experimental aid to help cool gases like oxygen and nitrogen, to such a temperature that they would liquefy. Commercial use to keep liquids hot was a successful by-product of this superb product.

 Zipper or zip 1893

Now a ubiquitous device for temporarily linking two edges of material together, the zipper or zip was conceived and invented by Chicago based mechanical engineer, Whitcomb L Judson. He devised the hook and eye principal of the zipper originally for shoes. Judson was granted patents for variations of his design in 1891, 1894 and 1905.

The Industrial Revolution 1700–1900

Great Britain, underpinned by coal, steel, iron and steam became the first and leading nation in the global industrial revolution. It is often said that Britain's Industrial Revolution created the modern world. It all began in Britain in the latter part of the 1700s, when most manufacturing was a cottage industry and essential products such as nails; clothing, timber and other basic produce of the day were produced by hand or with basic machines in people's homes. Even farming was a hugely manual and laborious task and very inefficient in its output. But rapid industrialization marked a surge towards purpose made powered machinery and specialist factories designed solely for mass production. The key driver behind this rapid development was the textile and iron industries that played a pivotal role in the rapid growth, innovation and impetus in the Industrial Revolution.

Supporting industries played a key role in the success of the Industrial Revolution, seeing huge developments in banking, Communication and transport systems. It was clear that such a huge cultural step change could not be achieved if the nation's people did not benefit. For the first time in Great Britain's history, this rapid industrialization not only generated considerable variety of manufactured goods to the people, but it also provided improved living standards for many, taking people from abject poverty and improving living conditions for the working classes and the poor.

Prior to the Industrial Revolution, the majority of people lived in very small rural communities, centered largely on farming. Life was not easy as low incomes meant that many people had to make their own clothes, furniture, tools and grow their own food. But this was all set to change. Industrialization attracted workers from rural communities with regular, higher wages, farming became mechanized, so fewer workers were required on the land and small towns centered around factories and mills were starting to grow at unprecedented speed into huge industrialized cities.

A number of factors contributed to Great Britain's role as the founder and birthplace of the industrial revolution. Firstly, it had huge deposits of iron ore and coal, essential for industrialization and British society was politically stable. Secondly, Great Britain was the most powerful colonial power at the time. This meant that it had an instant market for its manufactured goods and its colonies could also provide much-needed raw materials to power the enormous growth and demand in industrialization.

The rise of modernization and mechanization was fueled by the demand for British goods. A generation of innovation and invention developed through necessity as industrialization spread across the land. Eventually, other countries witnessed the growth in wealth and saw how industrialization could benefit them too. So the British introduced laws prohibiting the export of skilled workers and their valuable technology. However, these laws could not prevent the spread of industrialization to other countries, particularly, Germany, France, Holland and the United States of America.

By the mid 1800s, there was a well-established industrial base throughout the North-Eastern seaboard of the USA and most of Western Europe. Such was the speed of industrialization in other countries that Great Britain lost its place in the rankings of industrialized nations by the 1920's and the USA took its crown.

The Industrial Revolution

 ## Manufacturing revolution   1800c

Through the course of history, the process of manufacturing has developed spectacularly. Producers of goods have created methods to mechanize their systems and as a consequence reduce labor costs and speed production in the process. This transformation in production began in 18th century Great Britain and eventually spread to mainland Europe.

By the late 18th Century it had spread to the shores of America. The main benefactors of these huge changes were the industries of agriculture, mining, goods manufacturing, glass making, military hardware and textiles. At the height of the industrial revolution, Britain was the largest manufacturer by volume, in the world. Despite the small size of the country, Britain is still the world's 6th largest manufacturer.

 ## Sewage systems   1865

As the chief engineer in the mid 1800's for the Metropolitan Board of Works (LMBW) in London, Joseph Bazalgette made a huge impact on the health and wellbeing of the population of the capital through his innovative and efficient development of its large bore underground sewage system. Headed by Bazalgette, the establishment of the LMBW meant that for the first time in the city's history, a single organisation was able to plan and co-ordinate public civil works in a holistic way across the whole city. An outbreak of cholera prompted the city heads to power forward with a plan to begin works on street drainage and sewer improvements.

By the end of 1866 most of London was connected to a state-of-the-art deep sewer system that is still being used to this day. Bazalgette ensured that the new system of low-level sewers diverted all foul water and connected new and existing sewers to a network of riverfront treatment works. Thus, ensuring for the first time in the city's history, that no raw or untreated sewage was pumped into the River Thames. His legacy and inspiration lived on long after his death in 1891, with places as far afield as Mauritius, Port Louis and Budapest enjoying the benefits of his advice, design and planning.

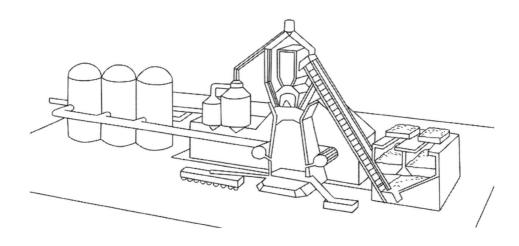

Industry Process and Innovation

 Aluminum - discovery   **1808**

Humphrey Davy identified the existence of the base metal alum, which he first named aluminum (an old English term still used in the USA) but later he renamed it aluminium. 17 years later the metal discovered by Davy was first produced in a very basic and impure form by Danish chemist Hans Christian Orsted.

 Barbed wire   **1867**

A type of fencing wire that is produced with sharp edges, points or wires that are arranged at regular intervals along the strands, making barbed wire the first choice for security fencing to this day. The person credited with this invention is Lucien B Smith from Kent, Ohio, who was granted a patent for it later in 1867.

Industry Process and innovation

 ### Battery - structural   2006
British company BAE systems developed an innovative method of building batteries into vehicle or body armor that also acts as an integral element of the structure. Designed initially to lighten a soldier's load on the battlefield, this technology is now used in many electric vehicle designs

 ### Bessemer converter - mass production of steel   1850
In the 17th and early 18th century, steel was very expensive to produce, using very inefficient power-hungry methods. However, in 1850, British inventor Sir Henry Bessemer was working on finding an inexpensive process for the mass production of steel. The key part of the process is to remove all impurities from iron by air oxidation blown through the iron in its molten state. This process in turn raised the temperature of the mass of iron, maintaining its molten state – now known as 'the Bessemer process'.

American, William Kelly around 1851, also independently improved on this process but after filing for bankruptcy, he was forced to sell his patents to Henry Bessemer. This additional information helped Bessemer fast track and refine a similar process that Kelly had also been working on, culminating in Bessemer's patent granted in 1855, for what was now a fully tried and tested process that was to change the world in steel production and use for the masses.

 ### Blast furnace - world's first coke fired   1709
Abraham Darby was an iron ore smelter who pioneered the use of coke to use in his blast furnaces. Because coke had fewer impurities, the iron Darby produced was of superior quality to iron used in a standard coal fired blast furnace.

Carbon fibre 1963

This fantastic super strong, lightweight material is only one of many inventions that were developed by British military researchers. Today Carbon Fibre has numerous applications ranging from cars, boats, sports equipment, record player pick up arms, bikes and even aircraft wings and fuselages, such as the Airbus A380 and the Boeing Dreamliner.

 1824

Joseph Aspdin was a cement manufacturer who was granted a patent for Portland cement in October 1824. The patent was entitled 'an improvement in the mode of producing an artificial stone'. It was within the text of the patent that he used the phrase 'Portland cement', based on its constituent ingredients, Portland stone and oolitic limestone. Prior to his invention, the most common cement used in bricklaying was a simple limestone mix. His invention was quickly adopted, first in Great Britain, and then across the world. Today, Portland cement is still the standard used for construction the world over due to its flexibility in use and temperature tolerance.

 Concrete - quick-drying 1774

John Smeaton was a mechanical scientist from Leeds. He started life as an instrument maker but later progressed into windmill and water wheel design. But it was his work on building the Eddystone lighthouse for which he is most famous. Savage seas and fires had destroyed the mainly wooden structures of its predecessors. Smeaton had to find a more innovative solution that was durable and long lasting. After researching many materials for use as an effective mortar, resistant to the harsh elements, Smeaton eventually developed a quick-drying concrete, known then as 'hydraulic lime'. His proposal to use clay and limestone materials using his method is regarded as the world's first use of modern concrete in engineering. To complete his task of building the lighthouse on Eddystone Rocks, he had to create a crane that was capable of lifting very heavy weights safely above the height of the lighthouse.

Industry Process and innovation

 Concrete - reinforced   **1854**

British plasterer William Boutland Wilkinson was recognised as being the first person to develop the method of building reinforced concrete. Wilkinson built a small reinforced concrete cottage and reinforced the floor and the wall with wire rope and iron bars. This was the world's first structural reinforced concrete building.

 Electro plating   **1840**

George R. Elkington and his Brother Henry were the British inventors who first patented the process of electroplating in 1840 and exploited the process commercially. The Elkington's had a virtual monopoly for many years based on their patent for an inexpensive system of electroplating. It must be noted that the Elkington's patent was based on the British inventor John Wright's design, to which he sold the rights for his process to the Elkington brothers.

 Escalator   **1859**

Designed to transport people from one level of a building to another, the escalator is effectively a moving staircase. Its design is based on a motor driven chain that allow the treads to remain horizontal and connected steps that travel up or down on tracks. Invented by Nathan Ames of Saugus, Massachusetts, although his design called 'revolving stairs' was never made.

 Fiberglass   **1938**

The process of heating and stretching glass into fine fibers has been acknowledged over many hundreds of years. Although, combining these fine fibers with a resin mix for the manufacturing of textiles and structural applications is far more recent. Russell Games Slayter of the company Owens-Corning, invented a new process and thus commercialized the development into a new product known as fiberglass or glass fibre.

Float glass   1952

Alistair Pilkington is an inventor who was knighted for his groundbreaking invention of the float glass process. Acknowledged to be one of the most significant inventions of the 20th century, Pilkington's process is now the principal method used to produce glass worldwide for windows etc. Almost all glass used throughout the world today relies on the development of British Glass Engineer, Pilkington's development from 1952. It is pure coincidence that Pilkington was, at the time of his discovery, working for British company Pilkington Glass and not related to the founders.

The Float glass process was originally only able to make glass 6mm thick. It has now been refined to make thicknesses from 0.4mm up to 25mm. Pilkington discovered that when molten glass is poured constantly at 1000 degrees Centigrade into a shallow bath of molten tin, then the glass floats on the tin, spreads and forms a supremely level surface. After controlled cooling, the glass materializes as 'process polished' with virtually perfect surfaces. Over 40 manufacturers in 30 countries, producing more than a million Tons of glass a week have now licensed Pilkington's process.

 ### Hydraulic press   1795

British locksmith Joseph Bramah developed unpickable locks, but more importantly was an avid inventor. The most significant of all his inventions was the Hydraulic Press. This machine consisted of two piston cylinders, each with a different cross-sectional area and they were connected with a tube, filled with a light viscous fluid. They were connected in such a way that one piston caused the other to move at the same time. It is still one of the most significant machine design concepts used in the industrial world today. Hydraulic presses are needed for all manner of production, from tin cans to rubber processing and car bodies to plastic molding.

Industry Process and innovation

 ## Linoleum   1860
Legend has it that British inventor Frederick Walton went to his shed to get a tin of paint and upon opening the lid, he noticed that a thick flexible film of congealed linseed oil had formed on top of the fluid. This provided Walton with a great idea. From that point, he knew that linseed oil could be formed into a waterproof material and that if he glued a canvas type of backing, he could market and sell it as an 'off the shelf' floor covering. Once in production and now known as linoleum, this ubiquitous material was sold the world over to millions of homes and offices.

 ## Macadam - road surface   1820
Macadam is a method of road construction developed by British engineer John Loudon McAdam in 1820. Although his creation was not new or unique, it did simplify previous systems of road building, through the use and standardization of layers of small size aggregate or small stones, bound by a wet-mix quick drying cement material, laid on a thick bed of compressed hardcore. Not to be confused with the more advanced early 20th century invention patented by British inventor Hooley in 1901 for the 'Tarmac' road construction method now used extensively throughout the world.

Oil - first oil tycoon   1851
James Young was a British chemist who was fascinated when he saw oil seeping out of the shale-laden ground near his home in Bathgate. After months of experiments and research, he refined the crude oil and created what we now know as paraffin, later developing the paraffin lamp. Young also invented the world's first lubricants for industry. He had in fact not only developed the world's first shale fracking technique of oil extraction, but in the process, became the world's first oil tycoon. Young further refined his crude oil process to produce what we now know as gasoline or petrol.

Oil refinery   1851

British chemist James Young built the world's first oil refinery to support his newfound discovery of oil in the Bathgate area. Eventually producing several grades of refined crude oil, Young paved the way not only for the British oil industry, but also the world's oil industry in general. As a result, Young formed the world's first commercial oil company.

 ### Photosensitive glass   1937

Invented by David Stookey whilst working for the Corning Glass Works, photosensitive glass is clear with minute metallic particles that can be charged by short wave radiation such as ultra-violet light, to produce an image or darkening screen on the glass.

 ### Pipe wrench   1869

An adjustable wrench used for turning soft iron fittings and pipes, the pipe wrench was designed with adjustable jaws that allowed it to bite into the frame so that any forward torque on the jaws would tighten them together, allowing the dog teeth profile to dig in to the soft iron pipe or fittings. The inventor of the pipe wrench was Daniel C Stillson in 1869.

 ### Pneumatic tires - air inflated   1887

John Boyd Dunlop invented the first practical pneumatic tires. Although the invention was first made in 1845 when railway engineer Robert W Thomson patented the world's first Pneumatic tires, but at the time there was little demand for them. It was more than forty years later when Dunlop further developed pneumatic tires to prevent his son from getting headaches when riding his tricycle. Fortunately, this time his development occurred simultaneously as the new mass-market bicycle trend was growing.

Industry Process and innovation

 Rotary printing press   **1843**

Richard Hoe successfully devised and patented the rotary printing press in 1843. A Rotary printing press is a machine that uses images printed on a screen that is then curved around a drum or cylinder. Then by rolling the drum over stationary plates of inked type, the imprint on the drums screen will leave an impression on paper, eliminating the need to take impressions directly off the plates.

 Safety fuse   **1831**

William Bickford was a leather merchant and inventor from Devon. He is best known for his invention of the safety fuse in the field of explosives. His first design put the main explosive into a parchment sealed tube and then joined it with a smaller parchment cartridge holding gunpowder as the fuse. Bickford later thought of a clever improvement by winding rope strands around a core of gunpowder. He then varnished the rope to ensure a steady burn up the rope from its end, eventually reaching the explosive. Bickford's improved method of safety fuse is still used to this day.

 Sandpaper   **1833**

Sandpaper or glass paper was first manufactured in London by British company John Oakey & Sons, a manufacturer of polishing materials and sandpaper. In Oakey's early days as an apprentice piano maker, he was taught to make sandpaper by gluing powdered glass or sand on to paper. As the head of his own business now, Oakey was able to develop better processes and adhesives. This made it possible to mass produce a cost effective and high-quality abrasive paper of differing abrasive grades and set up a new manufacturing business in London in 1833 to start full scale production. He later developed 'wet & Dry' and 'emery paper' abrasives along with a compliment of furniture polishes, silversmith's soap and plate powder. He died in 1887 and his sons John and Joseph carried on the business, eventually taking the company public in 1893. Today, Oakey Abrasives is still one of the world's most respected and leading abrasives companies.

Shopping cart 1937

Constructed from a metal or plastic basket on wheels, Sylvan Goldman's invention was first used in his supermarket chain in Oklahoma City.

 Stainless steel 1913

It was in 1912 that British researcher and steelworker Henry Brearly was given a job by a small firearms producer, to develop a substance that would extend the life of their gun barrels. During his research, he discovered a method to produce corrosion resistant steel through experimenting with the addition of varying quantities of carbon and chromium to normal steel. Legend has it that he discarded some steel that he had been experimenting with and a few months later, discovered that it still looked as shiny and bright as the day he threw the steel scraps outside. At the time, he simply named this new steel as 'rustless steel'. He first applied for a US patent in 1915 and now, stainless steel is used in everything from buildings, ship making and cars to surgical instruments.

 Steam hammer 1839

British inventor James Nasmyth created the steam hammer; a revolutionary machine that helped power the progress of the Industrial Revolution. Nasmyth had a simple concept. A hammering block was inverted and hoisted by steam to a position above the metal to be fashioned and then steam in the block was released and fell. It was so advanced that both frequency of blows and the strength of impact could be controlled accurately. Nasmyth's invention allowed for much larger forgings using heavier metals. The steam hammer was a direct predecessor of the pile driver.

 Steam shovel 1835

Designed for lifting and moving heavy material such as soil or rock, the steam shovel was possibly the first motor powered shovel and was invented by William Otis who received a patent four years later in 1839.

Industry Process and innovation

 ## Tarmac   1901

Tarmac is an abbreviation of Tarmacadam and is based on the type of road surface developed by Engineer John Loudon McAdam in 1820. However, in 1901, British civil engineer Edgar Purnell Hooley patented the name and the advanced process of Tarmac using Tar, rather than cement to bind the aggregate on the road surface. Hooley then founded Tarmac PLC and it is now one of Britain's largest construction and building material groups. The advanced principle behind Hooley's development on road servicing is still the basis for all tar-based road surfacing the world over.

 ## Urinal   1866

A urinal is a special toilet designed for men and boys, specifically for urinating only. These devices are normally wall mounted with manual or automatic flushing and drainage built in to the bowls lowest point. The urinals inventor Andrew Rankin patented his design on March 27th, 1866.

 ## Water and sewerage systems   1840

William Lindley was a civil engineer born in London in 1808. Working initially with his father, then later on his own, he designed and installed the first sewage systems in Europe. Notably, Warsaw, Lodz and Prague and he designed and coordinated the water system in Baku. He died in London in May 1900.

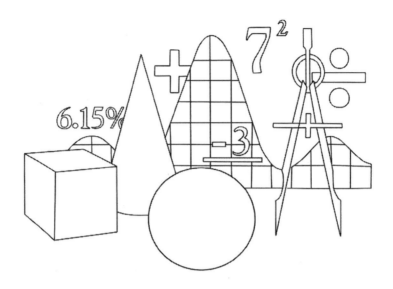

Mathematics

 Boolean algebra - digital logic design   **1840**

George Boole, a British mathematician, developed a generic technique for manipulating and representing logically valid inferences or 'mechanical ways of making transitions. Boole's approach used unequivocally, adopting algebraic systems for this purpose.

 **Calculus**   **1669**

Isaac Newton was one of two men who are independently attributed to the development of calculi and its foundations. Gottfried Wilhelm von Leibniz is the other great mathematician; although both arrived at similar conclusions, each man took a different path to reach their conceptual conclusions. Newton focused on Variables changing with time. Newton used quantities 'x' and 'y' that were limited velocities to compute the tangent. For Leibniz, the Calculus was taken towards analysis, while for Newton it was always geometrical. Nowadays, this subject is often taught conceptually first, in a rather reverse order compared to its actual development and as a result, today scholars are often taught first about its limitations.

Mathematics

 **Comptometer**   1887

Devised as a mechanical or electro-mechanical adding machine, the comptometer was activated solely by the action of pressing keys arranged in columns and was the first adding machine using this process. Invented by Dorr Felt, the device was capable of adding, subtracting, multiplication and division.

 **Equal sign**   1557

Mathematician and physicist from Tenby, Robert Recorde, introduced the plus sign (+) to English speakers and created the equals sign (=), both now used universally in the world of science.

 **Flow chart**   1921

Representing an algorithm, calculation or process, a flow chart represents an often complex mathematical calculations or outcomes by a graphical chart. Invented by Frank Gilbreth, flow charts are often used in displaying analysis, process flow and design and process flow charts.

 **Logarithms**   1614

British mathematician John Napier first discussed logarithms and published a paper of his findings in 1614 in a document entitled 'Mirifici Logarithmorum Canonis Descriptio' (Description of the Wonderful Rule of Logarithms). Napier is therefore recognised as the father and creator of logarithms.

Metric system   1668

Contrary to common belief that the French invented the metric system, it was in fact originally developed, with published findings by a British bishop called John Wilkins. Wilkins first published his system of decimal metrication, in a proposal named 'universal measure' in 1668. Wilkins developed his plan further that year and published another paper that soon became known as 'the metric system' after the Royal Society published 'An essay towards a real character and a philosophical language' and it contained quotes from Wilkins referring to his plan for a system that could be used as a 'single universal measure'. Wilkins later paper included all of the components of the international Metric unit-based System. The complete naming of all the metric units came much later, but there can be no doubt that his work lay the foundations for the adoption of the metric system in the modern world.

 ## Oil - the barrel measurement   1483

The global standard for the quantity of oil and the unit by which it is measured and traded is known as the 'barrel'. The barrel measurement dates back to the reign of the English king Richard III. Between 1483 and 1484, coopers had developed a method to make barrels watertight. During this period, a Cask held 84 gallons and a Tierce held 42 gallons. The Tierce or Barrel became the standard size for carrying anything including wine, beer and fish. The 42-gallon capacity of the barrel weighed about 300 pounds and was easier to maneuver than the much heavier cask. British oil pioneer James Young adopted this size and design. The American Petroleum Producers association formally adopted the 42-gallon Barrel in 1872 to hold and transport oil. The 42-gallon Barrel is now firmly established as the global standard measurement for oil.

   1854

Florence Nightingale was a gifted British mathematician as well as a famous nurse, who invented the 'coxcomb chart' now known as the pie chart.

   1622

William Oughtred was a British Anglican minister and recognised as the inventor of the slide-rule, the forerunner to the modern-day computer. In 1622 Oughtred placed two scales alongside each other and by sliding the scales back and forth to the desired point, he used them to calculate distance relationships, therefore dividing and multiplying directly without manual calculation. He is also credited with the invention of the circular slide rule. The slide rule was later used for roots, trigonometry and Logarithms by introducing a third sliding central scale.

Mathematics

 Standard deviation - probability 1868

In the 1840s, British mathematician Francis Galton, was fascinated with mechanical computing that performed both graphing and mathematical functions. Galton's greatest creation is probably his statistical correlation concepts and standard deviation. Galton was the first person to consider human differences such as intelligence, when collecting data, introducing what he called standard deviation into the output of data from questionnaires and surveys.

 Tabulating machine 1890

Used in accounting and to assist in summarizing data, the tabulating machine was a breakthrough for its inventor, Herman Hollerith. He devised a method of tabulating data results and calculations that were linked to a sorter or tabulator, that displayed the results on clock like dials. This was the start of automated data processing.

 Venn diagram - John Venn 1856

John Venn, who was an undergraduate student in mathematics at Gonville and Caius College at Cambridge University, developed the famous Venn diagram. Venn authored many books including 'symbolic logic' and it was during this period that he developed what is now commonly known as the Venn diagram. Venn's development was driven by what he considered the inadequacies of Euler's non-intersecting diagrams or Eulerian circles because they could not show inter-relationships across three differing classes of data.

"The study of mathematics, like the Nile, begins in minuteness but ends in magnificence"

Charles Caleb Colton

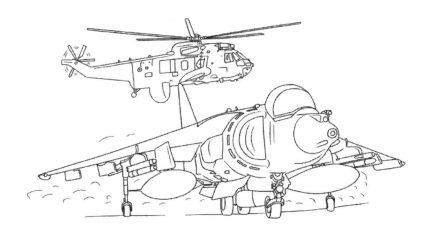

Military

 Aircraft carrier - HMS Argus　　　　　　　　　　**1918**

In this year, British naval architect James Graham's design of the world's first aircraft carrier was launched. Including all the major support systems and features such as: angled steel deck, steam catapult, tail hook, arresting cable system and aircraft command centre. It was the world's first carrier truly capable of launching and recovering naval designed aircraft.

 Bazooka　　　　　　　　　　　　　　　　　　**1942**

Designed as shoulder fired transportable recoilless rocket anti-tank firearm, the bazooka features a solid fuel rocket propulsion system, permitting high explosive anti-tank warheads to be targeted and delivered against machine gun encampments, armored vehicles and fortified bunkers, at far greater ranges than traditional grenades. The bazooka was co-invented by Colonel Leslie Skinner and Lt. Edward Uhl of the United States Army.

Military

 **Depth charge** 1914

Thomas Firth & Sons were a British engineering manufacturer from Sheffield. It is widely recognised that under instruction from the Royal Navy Torpedo school, they manufactured a depth charge based on Royal Naval Scientist, Herbert Taylors design, that went on to become the first useable depth charge in the world. It was designed in their words, 'for countermining as a dropping mine'. The mine was launched from a platform on the side of a warship and dropped into the water. The mines were fitted with a detonator that was actuated by water pressure automatically. If a submarine was in the vicinity, then mines of various depth setting could be launched to automatically explode at different levels. This proved hugely effective in forcing enemy submarines to surface quickly and surrender.

 Gas mask 1847

Worn over the face to protect the user from inhaling airborne contaminants, the gas mask formed a seal over the mouth and nose and also covered the eyes with 'windows' or built in goggles. Air was inhaled through a front facing bulb shaped filter and then vented through an exhaust in the rear of the mask back into the atmosphere. The gas mask was invented by Lewis Haslett in 1847 and he was granted a patent in June 1849.

 Dreadnought 1906

Dreadnought was the world's first turbine powered battleship, launched in 1906 and determined the design of all future big gun warships, dominating the navies of the world for the next 40 years. This mighty ship was 526 feet long with a crew of roughly eight hundred. Breaking with tradition, this behemoth boasted 4 propeller shafts driven by steam turbines rather than traditional pistons, providing a top speed of 21 knots. Aside from its standard above deck heavy armory, Dreadnought was built with 4 torpedo tubes to help fight off torpedo boats and destroyers. Never seeing action in battle before WW1, the Dreadnought was soon superseded by 'Super Dreadnoughts', with greater armament and better performance.

Fighter aircraft Vickers F.B.5.   **1914**

The Vickers F.B.5. Gunbus was a British designed and built fighter aircraft, first flown on July 17th, 1914, making it the world's first Fighter aircraft. The second crew member, or observer, sat in front of the pilot who was also responsible for aiming and firing the single 0.303 Lewis gun. It was the first aircraft specifically built for air-to-air combat.

 Jet bomber   **1949**

The world's first jet bomber was the English Electric Canberra. This revolutionary aircraft was also the world's first high altitude bomber.

It remained in service with several air forces around the world for more than 50 years and is still being used by NASA.

 Laser timing   **2008**

Used originally on the battlefield to prevent firing on friendly forces, this cutting-edge technology is now used in the Velodrome at Manchester to accurately time racers, designed and built by BAE (British Aerospace Engineering).

 Machine gun   **1718**

Inventor and lawyer from London James Puckle, is credited with designing and building the world's first machine gun: 'the Puckle gun'. Puckle demonstrated his tripod mounted single barrel gun, fitted with a nine-shot cylinder that automatically revolved after each round was fired. His weapon fired at a rate of nine shots per minute, contrasting with the standard musket at the time that could be loaded and fired at best, three times a minute. Unfortunately, although well documented and demonstrated, Puckle's ground breaking weapon failed to attract investment, prompting one newspaper at the time to comment; 'those are only wounded who hold shares within'. Puckle was the first person in the world to patent what was clearly described as a 'portable machine gun".

Machine gun – automatic　　　　　　　　　　　　　　1884

The machine gun today is described as a fully automatic firearm, capable of firing cartridges or bullets in succession from a magazine or belt. The world's first automatic machine gun was invented in 1718 by British inventor James Puckle and was capable of firing 9 rounds automatically in succession. But the more advanced machine gun invented by Hiram Stevens Maxim, was designed with the ability to fire up to 750 rounds per minute, without reloading a barrel.

 ## Magnetic proximity fuze　　　　　　　　　　　　　　1943

In 1943, inventor Panayiotis John Eliomarkakis of Philadelphia filed a patent for a proximity device that initiated a detonator when the fuzes magnetic equilibrium was disrupted by a ferrous metal object such as tin or steel passes over it. Used in naval mines, depth charges and land mines, Eliomarkakis was granted a patent on January 13, 1948.

 ## Nuclear submarine　　　　　　　　　　　　　　　　1955

In 1955, the world's first nuclear submarine was launched, the USS Nautilus. This new breed of submarine transformed naval warfare due to its nuclear powerplant that generated electricity in near silence, it needed no air for its engines and thus could stay submerged for months on end without the need to resurface for air. The nuclear submarine was invented by Admiral Hyman Rickover who led a team of scientists and engineers at the Naval reactors Branch of the US Atomic Energy Commission in 1951. The USS Nautilus became fully operational after 2 years of sea trials in January 1955.

 ## Periscope - collapsible　　　　　　　　　　　　　　　1902

An instrument used for observation from a hidden position, normally underwater and used mainly on submarines. The periscope is fundamentally a tube and at each end there is a mirror angled at 45 degrees, that allows the viewer to see 360 degrees by turning the periscope around, enabling the user to search for enemy targets. When out of use, the periscope can retract into the submarines hull. The invention of the collapsible periscope is credited to Simon Lake who called his creation the 'omniscope'.

RADAR - radio locator   1935

Sir Robert Watson-Watt designed and built the world's first RADAR capable of detecting targets and providing their true range. Watson-Watt was a British scientist who after graduating from University with a BSc, worked with Professor William Peddie. As a result of his work with Peddie, he became interested in radio waves and their application. In 1915 during WWI, Watson-Watt worked as a meteorologist for the Royal Aircraft Factory in 1915 and experimented with the use of radio waves to detect severe weather conditions to forewarn pilots of the perils that could lie ahead. In 1927 when Watson-Watt joined the National Physics Laboratory, he became an internal Superintendent of the unit.

In 1935, following many years of development and experimenting, he published a paper entitled "The detection of aircraft by radio methods". His report fascinated Sir Henry Tizard who was the head of the air defence committee. So much so, that he asked Watson-Watt to carry out what was to become a successful trial, in which short-wave radio was employed to detect an RAF bomber. Following the successful trials, Robert Watson-Watt was appointed as Superintendent of the new Air Ministry, Bawdsey research station near Felixstowe.

The ensuing research carried out by Watson-Watt and his team led to a network of RADAR stations being set up across the east and south coasts of England, proving vital in the defence of the country during the Battle of Britain. As a consequence, and for the first time ever, the RAF was given advanced warning of incoming attacks by the Luftwaffe, enabling them to prepare and react appropriately.

Watson-Watt and his team of John Randall and Henry Boot then advanced their research and developed a 'Cavity Magnetron'. This meant for the first time, a portable unit could be sited on a fighter or Bomber aircraft, giving the pilots even greater warning of the approaching enemy. Magnetrons could also heat up water and today, the same devices are to be found in every microwave in the world and used as a means of quickly heating foods or liquids. Watson-Watt was knighted in 1942 and given £50,000 for services to his country. The USA also presented him with the US Medal of Merit for his huge contribution in WWII. He died in 1973. RADAR stands for RADio Detection And Ranging.

Military

 RADAR - HS2 Airborne radar   1939

Alan Blumlein was possibly one of the world's greatest electrical engineers and pioneers. This British genius invented binaural sound form a single source (Stereo as we now know it), which he patented in 1933, then in 1936 he created an HD television system that would replace Baird's crude mechanical system. But possibly his greatest contribution to the world was his invention of Air Interception (AI) Radar. Alan Blumlein was instrumental in creating and implementing the world's first operational Air Interception Radar system. This huge technological advance played a key role in helping the RAF fend off the Luftwaffe night bomber offensive.

The most important Radar project in the latter part of the war was the H2S, a technique that enabled the RAF to accurately bomb through thick cloud. However, before he was able to see the fruits of his years of development and invention come to bear, he was tragically killed when a Halifax bomber that he was travelling in to test the system, crashed in the Wye Valley, killing all 11 crew including Blumlein. His teams of engineers were able to complete this crucial project successfully and it proved decisive in shaping the outcome of the war. He died aged 38 and to this day has received no recognition or honour for his huge contribution to science and the war effort.

 Rocket - liquid fuel   1926

Using a liquid propellant as fuel, Dr Robert H Goddard successfully launched the first liquid fueled rocket using gasoline and liquid oxygen for propulsion.

 Rocket - solid fuel   1804

British inventor William Congreve designed and built the 'Congreve Rocket', a British military weapon. The weapon was developed for the Royal arsenal following several bloody wars fought between the Kingdom of Mysore in India and the British East India Company. The rockets were the first of their kind to be powered by solid fuel and were used successfully in the Napoleonic wars of 1812.

Semi-automatic shotgun   1898

A self-loading shotgun is a firearm requiring only a trigger pull for each round, known in its original form as a semi-automatic shotgun. Its inventor John Moses Browning created the first weapon of its kind and its design dominated its niche sector in firearms for many years until a variant, the Browning Auto-5 stopped production in 1999.

Special forces - SAS and SBS   1941

The SAS and SBS are the world's first and leading Special Forces and were the model for the USA's special forces 'Delta Force'. SAS (Special Air Service) and SBS (Special Boat Service) were conceived and established in the early 1940s as elite regiments that could work and infiltrate behind enemy lines without detection. Admission to both elite units is by invitation only. Made up of mainly British soldiers who have to pass an exhausting selection test for acceptance and entry into the elite regiments. The SAS has become a template for most other countries' special forces. The SBS is the least famous, but equivalent of the SAS in the Royal Marines.

Ship - steam turbine   1897

Turbinea was launched as the world's first steam turbine powered ship in 1897. Built as an experimental Royal Naval ship it was at the time, the fastest ship in the world, setting the benchmark for years to come.

Sonar   1912

British mathematician Lewis Fry-Richardson was from Newcastle-upon-Tyne and created the world's first Sonar System. Sonar is the acronym for SOund Navigation And Ranging. Sonar works similarly to Radar but emits sound waves through the water and then bounces them back from submerged objects that lie hidden to the naked eye. Other countries have reproduced Fry-Richardson's invention. As reparation for WWII, Britain gave the technology to the Americans.

Military

 ## Stun grenades SAS   1960s
Britain's elite special service military unit, the SAS, developed the stun grenade. It sent a specification to the Royal Ordnance unit in Enfield North London and resulted in the 'G60 grenade' or 'flash bang' stun grenade. The stun grenade is designed to distract and disorientate the enemy in a building, providing the assault team critical seconds advantage to enter and counteract the threat. The SAS stun grenade was designed to create a powerful and loud bang exceeding 170 decibels in volume, with a blinding light flash and no lasting injury. Military forces worldwide have now implemented the stun grenade; such is its overwhelming success.

 ## Tank - military   1914
It was whilst reporting as an official war correspondent, that Ernest Swinton devised a design for a Military tank. He suggested that if the crawler tractors that were then used to pull military armory into battle on the Western front, were combined as a tracked cannon, then it would solve many issues, scoring a competitive advantage over the enemy. His design went into manufacture eventually in 1914 and it was the first 'Military Tank' in the world, capable of scaling a five-foot high obstacle and traversing six-foot trenches with ease. It was also armored to protect the crew and travelled at a leisurely 4 mph.

 ## Spar torpedo   1864
Consisting of a bomb positioned at the end of an extended pole or spar and attached to a boat, the spar torpedo was operated by guiding the spars end into an enemy ship. Although not technically a torpedo by todays terms, they were often armed with speared ends that would lodge in wooden hulls and detonated manually. The spar torpedo was invented by E C Singer who was employed by the Confederate States of America attached to a secret project's division.

 **Submarine**   **1578**

Designed by British mathematician and naval writer William Bourne, but not built until 1624 by Dutchman Cornelius Drebbel. Bourne is credited with designing and publishing a paper on the first credible submarine ship designed to be navigated underwater, distributed in 1578. Bourne put forward drawings and calculations for a totally enclosed 'underwater boat' that could be immersed and rowed underwater. It was built with a wooden frame and the body was covered in a thick skin of water-resistant leather. Its design enabled it to be submerged by contracting the boats volume through hand levers inside and then reversing the function to climb back to the surface.

 **Torpedo**   **1866**

British engineer, Robert Whitehead first designed and developed an underwater missile that could be launched from a ship in an underwater tube whilst at sea. His first version of 'Torpedo' was powered by compressed air rather than explosive charges and was fitted with a mechanism inside that enabled the missile to adjust and stay at a constant depth based on water pressure. It is on record that the very first ship to be sunk by Whitehead's torpedo design was a Turkish steamer called Initibah and was launched from a Russian warship in 1878. His invention notably changed warfare at sea.

"Who dares, wins.

Who sweats, wins.

Who plans, wins."

British Special Air Service (SAS)

Music and Musical Instruments

 **The Beatles**   **1960 to 1970**

The most successful pop group in the world. The 'Fab Four' as they were affectionately known, came from Liverpool and were formed in 1960. The four original band members were George Harrison, John Lennon, Paul McCartney and Pete Best. Ringo Starr replaced Best in 1962. Their manager was Brian Epstein, who encouraged George Martin to sign them up to the EMI owned Parlophone record label.

Their sales momentum and popularity did not diminish. The Beatles sold a staggering total of more than 600 Million records during the 60's alone, clocking up scores of top 10 hits and number ones, winning 14 international awards in that decade. Statistics alone prove that the Beatles were by far the first world super band and the most successful ever. Due to difficulties enforcing his patent, companies such as Sony and Philips, profited from his invention. By 1986, Sony and Philips and Time Warner were legally challenged and came to a licensing settlement and Time Warner was ordered to pay a patent infringement settlement of $30 million.

Compact disc   1970

An optical disc that is used to store data in digital form, originally designed for digital audio. The originator and inventor of the compact disc concept was James Russell who decided that the music industry needed a new medium involving no touching contact like the record player. Russell developed an optical system that used a low output laser light to read the digitally encoded data etched onto a plastic-coated metal disc.

Concertina   1829

Developed by Sir Charles Wheatstone and patented in 1829, he subsequently filed a second patent for an improved version in 1844. Probably developed separately, Carl Friedrich Uhlig announced a German version five years after Wheatstone's original patent. The concertina is a wind powered reed instrument, driven by hand held bellows and sound is generated by pushing and drawing, whilst simultaneously playing keys on either side of the hand-held bellow caps. Each end cap and bellow set is Hexagon shaped; though later variants were octagon shaped too.

Condenser microphone   1916

Also known as an electro-static or capacitor microphone, the condenser microphone comprised of a capacitor with two plates that has a potential between them. One of these plates acts as a diaphragm and consists of extremely light material. The diaphragm oscillates when hit by sound waves, varying the distance between the plates, thus changing the capacitance. Voltage to create this phenomenon is supplied by either a battery in the microphone body or by an remote external supply. The condenser microphone was invented at Bell Laboratories, by Edward Christopher Wente.

Music and Musical instruments

 ### 8-track cartridge   1964

Known as eight-track, the 8-track cartridge comprises a magnetic sound recording tape technology. Invented in 1964 by William Lear, the eight-track became one of the most popular music mediums from the late 60's to the late 80's. The tape was a continuous loop; thus, a whole album could effectively be played repeatedly until switched off, using four continuously playing stereo tracks that would switch from one to the other, hence the name 8-track.

 ### Elvis Presley   1953-1977

The uncontested best-selling solo artist in the history of recorded music, Elvis Presley received three Grammys and has been inducted into the multiple music halls of fame. Presley was successful across many musical genres including country, gospel, blues and pop.

 ### English horn   1720

The English horn dates back to 1720 and is a variant of the Oboe, being roughly one and a half times longer with a pear-shaped bell and pitched in F, rather than the standard oboe's C pitch. The English horn is very popular in France and is known there as the Cor Anglais.

 ### Glastonbury - largest music festival   1970

Created by British farmer's son Michael Eavis' with the inaugural Glastonbury music festival held in 1970, formerly known as the Pilton festival, this multi genre festival is by far the world's largest and the most famous music festival. Glastonbury has been running continually since 1970 with the exception of cyclical 'fallow years' to allow the land to recuperate. The festival's first headline act was Tyrannosaurus Rex, better known as T. Rex, and now the festival attracts music fans from all over the world, with tickets normally selling out within the first hour of release. The Glastonbury festival regularly attracts more than 185,000 festival goers with past headline acts such as Stevie Wonder, David Bowie, Rolling Stones, Dolly Parton, Muse, Adele and Coldplay.

Guitar – bass   1936

The bass guitar is similar to a standard electric guitar, with the exception that they generally only have 4 bass strings and typically, a longer neck. In 1936, inventor Paul Tutmarc of Seattle created the earliest solid body electric bass guitar, though not made from hardwood

 ### Guitar – hardwood solid body electric   1941

The solid body electric guitar now used by almost every rock band in the world, is typically made from a hardwood with a lacquer coating, having no internal open space to amplify vibration of the strings. Instead, the sound is only audible through electronic pick-ups on the guitar heads, that are positioned under the strings and on to an electronic amplifier and loudspeaker. Although not the first, the man credited with advancing the art of making this great instrument was renowned American musician, Les Paul.

 ### Gustav Holst   1874 to 1934

Born in Cheltenham, England, Gustav Holst is without a doubt one of Britain's greatest ever composers of the classical genre. His most famous composition is the Planets Suite. He became friendly with composer Ralph Vaughan Williams in 1895, the start of a lifelong companionship. In 1905 he became musical Director at St Paul's Girls School in Hammersmith and continued to teach there until the end of his life. It took Holst two years to write his most prolific works, the Planet's, between 1914 and 1916.

Holst received perhaps the greatest accolade of his illustrious career in 1929. He was awarded the Howland Memorial Prize from Yale University for excellence in the arts and the gold medal of the Royal Philharmonic Society in 1930. He was appointed visiting lecturer at Harvard University in 1932 but became ill, forcing him to return to Britain, later dying on 25th May 1934.

Music and Musical instruments

John Williams   1952-

A renowned composer, pianist and conductor, John Williams is recognised as one of the greatest American composers of all time. In a career spanning more than six decades, he has composed some of the most recognizable and critically acclaimed film scores in cinematic history.

Jukebox   1927

Generally designed as a coin operated automatic music playing machine, the multi-play juke box was first conceived and invented by 'The Automatic Music Instrument Company' who coined the name 'Jukebox'. The machine enabled a user to insert a coin and select a song from a selection of shellac made 78 rpm records, later to become ubiquitous with the advent of the more practical 7-inch vinyl single play records in the 1950's.

Live aid   1985

Live aid was a concert born in Britain as a response from the natural disasters sweeping Africa at the time and the subsequent need to raise money to aid immediate disaster relief. It was conceived by Midge Ure, Harvey Goldsmith and Bob Geldof and after only twelve weeks of planning, evolved as the most successful one-time charity fundraiser of all time, raising in excess of one hundred and fifty million pounds. It is also recognised as being the world's largest and most successful music concert, seen by more than two billion people worldwide. In attendance across both venues were fifty-eight of the world's most prolific bands and pop artists giving their time for free. More than one hundred and seventy thousand people attended the two venues in London and Philadelphia USA. Phil Collins famously became the only act in Live Aid to play at both venues on the same day, by performing and finishing his set in Wembley Stadium, London, then jetting off to JFK, New York by Concorde to play the same song set in Philadelphia later that day.

Moog synthesizer 1964

An analog synthesizer that doesn't use a vacuum tube, the Moog synthesizer employs analog circuits and computer methods to produce electronic sound. The Moog synthesizer was invented in 1964 by Dr Robert Moog.

 Synthesizer 1876

An electronic instrument capable of creating a myriad of tones and sounds by generating electronic signals of differing frequencies, the synthesizer was a versatile device. Elisha Gray was credited with this pioneering phenomenon. Gray later developed a basic loudspeaker in later variants to make the synthesizer audible.

 Theatre organ 1887

The theatre organ is a unique air powered pipe organ and is often called the cinema organ. Robert Hope-Jones is considered to be the inventor of the theatre organ, conceived from a desire to simulate the entire ensemble of orchestral instruments in one device. He built 246 complete organs between 1887 and 1911 and eventually merged his business with the famous US Wurlitzer Company in 1914.

 Tuning fork 1711

Still used today, by professional piano tuners in particular, to tune genuine stringed pianos. British musician John Shore invented a finely tuned steel fork, that when tapped with another steel object would emit a perfect pitch in a particular key through resonance for several seconds, thus allowing a piano tuner to set for instance middle C as the benchmark for all other strings. Prior to his invention, tuners relied on pitch pipes made of wood and were susceptible to pitch changes due to ambient temperature and humidity changes. The tuning fork had none of these negative characteristics, instead holding its pitch perfectly across a whole range of environmental conditions. The tuning fork is still in use today, although as in many areas of science, electronic devices have now replaced it.

Music and Musical instruments

Yesterday - The Beatles   1965

The world's most covered song is Yesterday by The Beatles. According to the Guinness World Records and has been covered seven million times in the 20th century alone. Versions range from Frank Sinatra to Wet Wet Wet and in more than 100 different languages. It also holds the title for being the most played record on radio in the world.

TOP 10 British and American bands & singers by sales

#	Artist	Sales
1.	The Beatles	178m
2.	Garth Brooks	148m
3.	Elvis Presley	250m
4.	Led Zeppelin	112m
5.	Eagles	101m
6.	Michael Jackson	81m
7.	Elton John	78m
8.	Pink Floyd	75m
9.	George Strait	69m
10.	The Rolling Stones	67m

(Source RIAA) - Platinum and gold record sales only

"If music be the food of love... play on."

William Shakespeare

Photography and cinematics

 Camera - multiplane   **1933**

Invented by Ub Iwerks, a former Disney animator, the multiplane camera was a special movie camera used in the conventional animation process. It transported a number of artwork pieces at varying velocities and at different spaces between one another, in front of the camera's lens delivering a three-dimensional effect. The key to success was to ensure that movement is calculated and shot frame by frame to present the impression of depth, by several layers of film moving at differing speed.

 Cartoon - first moving cartoon film   **1900**

J Stuart Blackton was a British film producer. After travelling to the USA, he is recorded as being the first person to sketch and direct the world's first cartoon film called 'The enchanted drawing'. This was a ground breaking silent cartoon movie, famous for containing the first animated sequences recorded on standard picture film.

Cinematography - creator   1889

William Friese Greene was a portrait photographer and inventor. He built and operated studios in both Bristol and Bath and later added two more, one in Brighton and the other in London. He met John Rudge who had developed a 'Magic Lantern' that was unique in the way it could display seven slides in swift progression, giving the effect of movement. Friese-Greene was enthralled by what he had seen and was soon working with Rudge to develop his Lantern so that it could project photographic plate images on to a wall. They named this device a 'Biophatascope' and realizing that plates had huge limitations, started experimenting with oiled paper and then progressing to celluloid film.

On June 21st, 1889 Friese-Greene was issued with a patent for his new 'Chronophotographic' Camera. Capable of shooting 10 frames per second using perforated celluloid to enable a cog either side of the film, to turn the reel of film by a handle. However, due to the low frame speed and the unreliability of the device it did not gain the popularity he had hoped. He later experimented with stereoscopic moving images and later in the 1890s, colour moving images. His 'hobby' of cinemagraphics had unfortunately drained funds from his other businesses and he was officially declared bankrupt in 1891. Friese-Greene was widely regarded as being 'the father of cinematography'.

Digital camera   1975

By digitally recording images through an electronic image sensor, the digital camera has revolutionized the world of photography and even mobile smart phones. Kodak Eastman engineer, Steven Sasson is credited with inventing and building the first digital camera using a CCD image sensor in 1975.

Instant camera   1923

A variant of camera that used self-contained, self-developing film, the instant camera was invented by Samuel Shlafrock. It consisted of a portable darkroom and a camera combined into a single box.

Kinemacolour - colour movie process   **1906**

British inventor George Albert Smith invented Kinemacolour. It was the world's first commercially successful colour motion picture system and started to be used commercially in 1908. The Kinemacolour process was effectively unchallenged until 1914. Smith came from Brighton and was hugely influenced by the day's great pioneers such as Edward R Turner and William N L Davidson during the process's development. Charles Urban of Urban's Trading Company in London, marketed Kinemacolour on Smith's behalf.

From 1909 it was rebranded as Kinemacolor, because Urban thought the name would be better accepted in lucrative American market. Kinemacolor used a two-color additive process, by filming and projecting a black and white film behind alternating green and red filters. The film had to be shot at 32 frames per minute rather than the standard 16fpm and was fitted with a spinning colour filter as well as a standard shutter. The filter consisted of four equal sections, two filled with green dyed gelatin two filled with red dyed gelatin, so that the four colour combination was red, green, red, green. Panchromatic film was used and the negative processed normally. There was actually no colour in the film itself.

The first successful screening using Kinemacolour was called 'A visit to the seaside'. It was shown in Brighton in September 1908. Such was its success, that more than 300 cinemas throughout Great Britain installed Kinemacolour projectors. More than three hundred major films were shot and filmed using the Kinemacolour process.

 Moviola   **1924**

A device that provides the film editor with the ability to view film whilst editing, the moviola was the first device in movie editing that enabled the editor to study individual shots to determine the best edit cut-point. General film editing used a vertical moviola and became standard throughout Western movie making until the 1970's. The Moviola machine was invented by Dutch-American Iwan Serrurier.

Photographic film 1885

Generally, a sheet of material that is coated with a photo-sensitive emulsion, photographic film was invented and developed by George Eastman and his company, Eastman Kodak. If the emulsion is exposed to light or other forms of radiation, it forms an image. Before photographic film was invented, this process was carried out using photographic plates.

Photographic plate - mechanical production of 1879

The forerunner of photographic film, photographic plates consisted of light sensitive mixture of silver salts applied to a glass plate. Although this form of photographic material eventually disappeared from the consumer market as less fragile film appeared. The inventor of the photographic plate was George Eastman who was granted a patent on April 13th, 1880.

Photography - practical 1835

It is very difficult to say exactly who invented photography. Certainly in 1826, we know that Frenchman Joseph Niepce created the first fixed image that took more than 8 hours to expose. However, what made British inventor William Henry Fox Talbot's invention ground breaking in the world of photography, was that by coating paper with silver oxide he had discovered a method of producing a translucent negative that could be used again and again to make any number of 'positive' photographs through contact printing. This became a common method used from this point on until the more recent arrival of digital photography.

Stereoscope - 3D 1838

Sir Charles Wheatstone was a physicist, born in Gloucester in 1802. Wheatstone developed a device for observing pictures in three dimensions. He called the instrument a mirror stereoscopic viewer that required two pictures of the same subject to be reversed laterally. The benefit of this arrangement was that it was not just limited to small images. This is why his stereoscope principle is still in use today for screening X-Ray 3D pictures and aerial images

Publishing and authors

 Agatha Christie   1890-1976

Agatha Christie was a British crime novelist and playwright. Guinness Book of Records notes that she was and still is the best-selling novelist of all time, with her novels selling more than 4 billion copies worldwide. She also wrote under the pseudonym of Mary Westmacott and was credited with six romantic novels. Born into a wealthy middle-class family in Torquay, she is best known for her 66 crime novels and 14 short stories written under her own name.

Most of Christie's crime novels revolve around the characters of Hercule Poirot and Miss Jane Marple. Finally, she is also credited with writing the world's longest running play; 'the Mousetrap' that first opened its curtains in 1952 and is still running to date. Its 50th anniversary was held on 25th November 2002, whilst its 25,000th performance was held on 18th November 2012.

Publishing and authors

 **Anne Brontë** 1820-1849

A British novelist, Anne Brontë was the youngest daughter of the literary Brontë family. She spent most of her life in the Yorkshire moors, living with her family. After leaving boarding school at the age of nineteen she went on to become a School governess and then a teacher. After her teaching career, she became a writer of poems and novels. Her most notable novels being Agnes Grey and The Tenant of Wildfell Hall.

 Arthur Conan Doyle 1859-1930

Sir Arthur Conan Doyle was a British novelist who is mainly famous for his many books on the fictional private detective, Sherlock Holmes.

 Arthur Miller 1915-2005

A playwright and a master of American theatre, Arthur Miller is regarded as one of Americas most significant writer of dramas, including 'The Crucible, All My Sons, A View From The Bridge and Death of a Salesman.

 Beatrix Potter 1866-1943

Born in Kensington, London, Potter was schooled by a private governess and had infrequent contact with other people. She loved drawing and studying her pets, becoming the inspiration for her many children's novels. Her parents often visited the Lake District for the three summer months of the year and rented Wray Castle, Ambleside. Over her many visits to Ambleside, Potter was inspired by her love of the countryside and the animals within, leading eventually to her writing and publishing her first book: The tale of Peter Rabbit in 1902. Other novels of note were: The Tale of Benjamin Bunny, The Tale of Squirrel Nutkin and The Tale of Jemima Puddle Duck.

 Cambridge University Press 1534

Operating continuously since 1584, Cambridge University Press (CUP) is the oldest publisher and printer in the world. CUP is the second largest University press in the world next to Oxford University Press, which is the world's largest.

Charles Dickens   1814–1870

Dickens was a prolific British author of some of the most creative and enduring literary classics. Dickens books are read throughout the world as staple English Literature, such as: The Old Curiosity Shop, A Christmas Carol, Oliver Twist, A Tale of Two Cities, Great Expectations & The Pickwick Papers.

 ### Charlotte Brontë   1816-1855

British poet and novelist, Charlotte Brontë, one of the famous three Brontë sisters and is noted for her brilliant novel's such as: Jayne Eyre, Shirley, The Professor and Villette.

 ### D.H. Lawrence   1815-1930

David Herbert Lawrence was a British author widely acknowledged in shaping 20th century fiction writing. His most famous and controversial novel was Lady Chatterley's Lover in 1928.

 ### Dictionary   1755

Dr Samuel Johnson was a British scholar who published the world's first English language dictionary, simply entitled 'A Dictionary of the English Language'. It is for this reason that it is regarded as one of the most influential dictionaries in the history of the English language. There were limited forms of dictionary produced before Johnson's publication, but Johnson's was the first to focus on words used in everyday conversation aimed specifically at the layman. It was not until 173 years later when the Oxford English Dictionary was published, that a competitor was of the same standard and detail as Johnson's original work.

 ### Earnest Hemmingway   1899-1961

A novelist, journalist and short story writer, Earnest Hemmingway published six short stories, two non-fiction books and seven novels, eventually winning the Nobel Prize for Literature.

 Edgar Allan Poe   1809-1849

A writer, literary critic and editor, Edgar Allan Poe is renowned for his short stories and poetry, predominantly his mystery and suspense tales. He is widely regarded as the inventor of American detective literature and regarded as the master of the short story.

 Emily Dickinson   1830-1886

Renowned as one of the most powerful writers of American culture, Emily Dickinson's work has enthused other writers such as the Bronte sisters.

 Emily Jane Brontë   1818-1848

A British poet and novelist and one of three famous literary sisters, Emily Brontë was born in Yorkshire in 1818. She is remembered mostly for her one and only novel: Wuthering Heights. This masterpiece is now recognized as a Classic piece of English Literature.

 E-paper   1973

Also called electronic paper, e-paper is a display device or technology, designed to simulate the appearance of normal ink on paper. The key benefit of e-paper is that is capable of maintaining a still image without drawing power from the mains or battery supply. Most common examples of e-paper technology are e-books and e-readers. E-paper was invented by Xerox engineer, Nick Sheridon, whilst working at the company's Palo Alto Research Centre.

Enid Blyton   1897-1968

Born in London, Enid Mary Blyton was a prolific children's writer whose books have sold more than 600 million copies. They include such classics as: The Famous Five, Noddy, The Secret Seven and The Enchanted Wood.

English book - first printed　　　　　　　　　　　　**1475**

The Recuyell of the Histories of Troye was the very first book to be printed in the English language by William Caxton, printed and published in 1475. It was translated by Caxton from French to English and took him almost 4 years to complete, due in part to his poor grasp of the French language but was later helped and encouraged by Margaret Duchess of Burgundy. After setting up his printing press in his Westminster home in London, he proceeded with the onerous task of translating and printing several other notable French works, all published in English.

 F Scott Fitzgerald　　　　　　　　　　　　**1896-1940**

American novelist, Francis Scott Fitzgerald is regarded as one of the greatest American writers of the 20th century. Fitzgerald is most famous for his epic novel, The Great Gatsby.

 George Byron　　　　　　　　　　　　**1788-1824**

George Gordon Byron was a British poet, who inherited his great Uncles title of 'Lord Byron' at the age of ten. Byron was famous for such books as: Don Juan, Manfred and The Giaour.

 George Elliot　　　　　　　　　　　　**1819-1880**

Mary Anne Evans was best known by her pen name as George Elliot. She was a Victorian British translator, journalist and novelist. George Elliot was best known for the novels: The Mill on the Floss, Middlemarch, Silas Marner and Daniel Deronda.

 George Orwell (Eric Arthur Blair)　　　　　　　　　　　　**1903-1950**

George Orwell was a British novelist born in Bengal, India to British parents who attended the world-famous Eton College. His most famous novels include, Animal Farm and Nineteen Eighty-Four.

Publishing and authors

 **H.G. Wells** **1866-1946**

Herbert George Wells was an English author, most famous for his science fiction writing. Notable works include; The Time Machine, The Invisible Man, War of the Worlds, The First Men and When the Sleeper Wakes. In the literary world, Wells was widely regarded as 'The father of science fiction'.

 **Henry James** **1843-1916**

Regarded as one of the key literary figures of the 19th century, Henry James writing style focused on writing from the characters perspective, allowing him to explore perception and consciousness. He was nominated three times for the Nobel prize for literature.

 **Herman Melville** **1819-1891**

Melville was a writer of short stories novels and poems and is most famous for his epic novel, Moby-Dick, often cited as America's greatest novel.

 **John Steinbeck** **1902-1968**

The author of sixteen novels and five short stories, John Steinbeck is most widely known for his masterpieces, Of Mice and Men, East of Eden and The Grapes of Wrath, which itself has sold more than fifteen million copies.

 **Joseph Heller** **1923-1999**

An American satirical writer, Joseph Heller was also renowned for his short stories and plays. He is most noted for the classic book and film, Catch 22 and ranks as one of Americas greatest novelists.

 **Kingsley Amis** **1922-1995**

Born in London, Kingsley Amis was a poet, teacher and novelist, best remembered for the novels: Only Two Can play, Lucky Jim, The Green Man and Take a Girl Like You.

Ian Fleming   1908-1964

Ian Fleming was a British author, journalist and naval intelligence officer. Most famous for his novels based on the fictional British spy; James Bond. Fleming sold more than 100 million James Bond books; the vast majority turned into worldwide film blockbusters. He also wrote the famous novel: Chitty-Chitty-Bang-Bang, later made into a major global film blockbuster and play.

 ### Jane Austen   1775-1817

Born in Steventon, Hampshire, Jane Austin was a novelist famous for romantic fiction based around the English upper class and is famous for such literary classics as: Pride and Prejudice, Becoming Jane, Sense and Sensibility and Mansfield Park.

 ### J.K Rowling   1965-

World famous author of the Harry Potter novels was born in Yate, near Bristol, UK. J.K. Rowling has sold more than 400 Million books worldwide and the film adaptation of Harry Potter became the world's most lucrative and successful film franchise in cinematic history.

 ### J.M. Barry   1860-1937

British author and dramatist J.M. Barry, is remembered most for his novel Peter Pan, later turned into a blockbuster movie by Disney. The name Wendy was first mentioned in this story and has gone on to be a hugely popular girls name ever since.

Publishing and authors

 Lancet - first medical journal 1823

With the first edition of 'The Lancet' published on 5th October 1823, the publication became the world's longest running medical journal, still produced in its thousands today. Thomas Wakley was a British Member of Parliament and medical coroner who started the Lancet publication in partnership with James Wardrop, William Lawrence and William Cobbett. By 1830 it boasted a circulation of more than 4000 copies. The Lancet was, at times, very controversial, often exposing bad practice in the medical profession, to such an extent that over its many years in publication, it is probably one of the single biggest drivers of improvements in the medical profession and its practice.

 Laser printer 1969

A common type of computer printer, the laser printer produces high quality text and graphics at speed, on plain paper. The laser printer was invented at Xerox in 1969 by researcher Gary Starkweather.

 **Mark Twain** 1835-1910

Samuel Langhorne Clemens is known more famously by his pen name, Mark Twain. Formerly a Mississippi riverboat pilot, he progressed to become an American literary legend. His most famous novel is Huckleberry Finn.

 Newspaper - first printed in Braille 1892

The first newspaper to be printed in Braille was the British publication 'The Weekly Summary'.

 Printing press - rotary roll feed 1865

The worlds earliest paper roll feed rotary printing press was invented by William Bullock in 1865. It was the first press that could continuously feed paper and print both sides simultaneously. Bullock's press became an American industry standard.

Roald Dahl   1916-1990

Born in Britain to Norwegian parents, Roald Dahl rose to the senior rank of Wing Commander in the Royal Air Force as a successful fighter pilot before embarking on his hugely successful career as one of the world's best-selling authors, in both the adult and children's genre. He was responsible for some of the biggest selling books in the twentieth century literary world, including: Charlie and the Chocolate Factory, Matilda, James and the Giant Peach, BFG, Tales of the Unexpected and Fantastic Mr Fox. All were turned into best-selling movies or TV dramas.

 ### Robert Louis Stevenson   1850-1894

Born in in Edinburgh, Robert Louis Stevenson was a poet and novelist most famous for his works: Treasure Island, Kidnapped, Strange Case of Dr Jekyll and Mr Hyde and The Master of Ballantrae.

 ### Rudyard Kipling   1865–1936

Rudyard Kipling was a British author born in Bombay (now Mumbai), India, to British parents and educated in Britain. Most famous for his novel The Jungle Book, more recently (20th Century) made into a blockbuster classic cartoon by the Disney Corporation.

 ### Shakespeare   1564-1616

William Shakespeare is recognised as being the world's greatest and most successful playwright and dramatist. Shakespeare was born in Stratford-upon-Avon in the English county of Warwickshire. His Father – John Shakespeare, was a local glove maker and wool merchant. His mother Mary Arden was the daughter of an upper-class landowner from Wilmcote.

Shakespeare was educated at nearby King Edward VI Grammar School in Stratford. He married young at 18 to the daughter of a local farmer, Anne Hathaway. She was 26 when they married and Susanna was born 6 months later after their wedding. They had twins, Judith & Hamnet two years later, but their son died aged 11 years.

Shakespeare started his professional theatre career as an actor and playwright and was one of the managing partners of the Lord Chamberlain's company, subsequently known as the Kings Company after James succeeded to the throne. The company bought two theatres in the Southbank area of London; the Globe and Blackfriars. Shakespeare's first published poems 'The rape of Lucrece' and 'Venus and Adonis' were published between 1593 and 1594. As a playwright, Shakespeare was prolific. His first plays began to materialize in 1594 and he continued to average two each year up to 1611.

He began to prosper soon after his first publication with his plays, mainly histories and comedies such as 'A Midsummer Night's Dream', 'Richard II', 'Romeo and Juliet', The Merchant of Venice' and 'Titus Andronicus'. Towards the end of Elizabeth I's reign, Shakespeare was well established as a famous poet and playwright. Moving from London, he spent the final years of his life in Stratford-upon-Avon.

Shakespeare passed away on 23rd April 1616 at the age of 52 and was buried at Holy Trinity Church in Stratford, having left a legacy of 37 plays in all, that are now taught in schools internationally as a staple in the subject of English Literature.

Shorthand   1837
British author, Sir Isaac Pitman developed a shorthand method of writing, in the phonetic system to become the main shorthand method in the English-speaking world. Symbols signify sounds not letters and words and by and large are written in shorthand as they are spoken. Other shorthand systems followed, but Pitman was the most popular in the UK and second in the USA.

Stephen King   1947-
Based on book sales alone, Stephen King has sold more than 350 million copies, some that have been adapted into box office movies. To date, King has penned two hundred short stories and fifty-eight novels and is often described as the king of horror.

Stereotyping 1725

William Ged was a goldsmith who invented a method in which a whole page of printing type could be cast as a single mold, better known as Stereotyping.

 Telephone directory 1886

Telephone directories are listings of subscribers within a given geographical area or service providers, in alphabetical order. R H Donnelly invented and created the first official directory and was referred to as the Yellow Pages in 1886.

 Tennessee Williams 1911-1983

Thomas Lanier Williams, better known as Tennessee Williams, was one of Americas greatest playwrights in the 20th century. Writing more than thirty plays, some classic western dramas, he also wrote short stories and novels.

 Thesaurus 1852

Peter Mark Roget was a British, London born lexicographer, natural theologian and physician who published his 'Thesaurus of English Words and phrases', better known as Roget's Thesaurus. Upon his retirement from medicine, Dr Roget started his life ambition of compiling a reference book that would contain the most used English Phrases and English Synonyms, with the aim of stimulating ideas and assisting in literary arrangement and composition. His book Roget's Thesaurus is still the most famous and continuously published book of Synonyms in the English-speaking world.

 Thomas Hardy 1832-1898

Born in Dorset, Thomas Hardy was a Victorian poet and romantic novelist. His most notable works include: Far From the Madding Crowd, The Mayor of Casterbridge, Jude the Obscure and The Return of the Native.

Publishing and authors

 Toni Morrison   1931-

Known for their detailed characters and vivid language, Toni Morrisons most famous novel is her 1987 book, Beloved. She was awarded the American Book Award, the Pulitzer Prize and the Nobel Prize for Literature.

Touch typing   1888

A touch typist will memorize the keyboard, knowing the keys location through muscle memory. Touch typing was invented by court stenographer Frank Edward McGurrin, from Salt Lake City, Utah.

 T S Eliot   1888-1965

An American born British poet, playwright, essayist and critic, Thomas Stearns Eliot is now recognised as one of the 20th centuries major writers and poets. Awarded more accolades than any other American born writer from the past two centuries, he has also received the Nobel prize for Literature and the British Order of Merit.

   1906

Described as the process of setting material in type or into a form to be used in printing for visual display, typesetting was invented by New Yorker, John Raphael Rogers. A patent was granted to Rogers on November 27th, 1906.

   1714

Henry Mill was granted the world's first patent for a typewriter. Mill was a British inventor who worked as an engineer at a local waterworks. The device he invented looked to be similar to a standard typewriter of the 20th century, although he described it as a 'machine for transcribing letter'. Nothing more is known, but it is certain that further development by engineers and inventors through the centuries used his designs as a basis for the modern typewriters of the 20th century.

Virginia Woolf 1882-1941

Born in London, Virginia Woolf was a novelist of the modernist genre. Her most famous works include: Mrs Dalloway, The Waves, A Room of One's Own, and To the Lighthouse.

 Wendy 1904

The first girl in the world to be named Wendy was Wendy Darling in the J.M. Barry play Peter Pan, later turned into a book in 1911. Since the play and book were published, Peter Pan has become a staple character in children's libraries.

"I don't think necessity is the mother of invention. Invention arises directly from idleness, possibly also from laziness. To save oneself trouble."

Agatha Christie

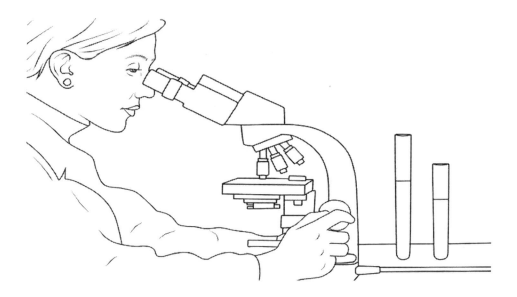

Science

 Atom - discovery of   **1911**

Ernest Rutherford was a New Zealand born British physicist, who through his ground-breaking research work attracted the unofficial title of the father of nuclear physics. Rutherford discovered the concept of radioactive half-life, by proving that radioactivity involved the metamorphosis of one chemical element to another. He also noted differentiation of atoms and named Alpha and Beta radiation. In 1907, he moved what is now known as the University of Manchester, where he worked with Thomas Royds, proving that alpha radiation is in fact helium ions.

In 1911, Rutherford made his most significant discovery where he theorized that atoms concentrate their charge in a very small core or nucleus, although he couldn't prove whether it was positive or negative at the time. This led to him establishing the Rutherford model of the atom, through 'Rutherford scattering' in his famous gold foil experiment. He is also widely recognised with the first 'splitting of the atom' in 1917 through a nuclear reaction between nitrogen and alpha particles, in which he discovered and named the proton.

Atomic theory 1805

British chemist John Dalton came a from a Quaker background, which may have contributed to his modesty that ultimately prevented him accepting fame and recognition for his pioneering work that we know today. He is known mainly for his work on atomic theory and although this work is now more than two centuries old, Dalton's theory still holds valid in modern chemistry and teachings to this day.

Dalton's theory states that:

1. All matter is made of atoms. Atoms are indivisible and indestructible.
2. All atoms of a given element are identical in mass and properties.
3. Compounds are formed by a combination of two or more different kinds of atoms.
4. Chemical reaction is a rearrangement of atoms.

Modern atomic theory is now of course a little more complex than Dalton's original theory, but the fundamental nature of it remains valid. The main advances in knowledge mean that we now know that atoms can be destroyed through nuclear reactions but not via chemical reactions. There are also differing atoms (by their masses) within each element and these are now known as isotopes. His theory though, rapidly became established as a theoretical basis in chemistry worldwide.

 Bifocals 1784

Benjamin Franklin, prolific inventor and past president, is credited with the invention of the first pair of bifocals in the early 1760's. Bifocals are unique in that they provide corrective lenses in two distinct regions within the whole lens's optical power. Typically, near sight in the lower half of the lens and longer distance in the upper half.

Science

✳ Cathode ray tube   1879

Sir William Crookes was a renowned British scientist, who led the way for many great discoveries. His development and research on a range of vacuum tubes was pivotal in his development of the cathode ray tube. On 22nd August 1879, he demonstrated several different vacuum tubes and discussed plasma, as 'the fourth state of matter'. Many of Crookes' discoveries based around his tubes laid the foundation for further discoveries such as the 'Braun tube' and the 'X-Ray' tube. The Braun tube developed later into the Television Tube.

✳ Clouds - categorization of   1802

British manufacturing chemist and amateur meteorologist from London, Luke Howard, was widely accepted to have created the categorization and put names to cloud formations, in his proposal and presentation to the Askesian society in 1802, now standardized and accepted worldwide. He observed through his research, that clouds were grouped into distinct types and that each cloud category had a different cause and effect on the atmospheric weather conditions. He noted that clouds were evident pointers into the operation of these causes. He was the first person to observe and classify each cloud type in universal Latin.

✳ Geological timescale   1931

Geologist Arthur Holmes first proposed a geological timescale, soon after the finding of radioactivity and its use. In his proposal, Holmes calculated that the Earth was roughly four billion years old, much older than had been previously thought.

Maritime clock - marine chronometer   1720's

For hundreds of years of maritime adventure and discovery, one problem always hampered accurate navigation at sea and that was known as the 'longitude problem'. For every 15 degrees that a person travels east, then the local time progresses one hour ahead and when travelling west, the problem is reversed.

John Harrison was a joiner from Lincolnshire with little formal education, who sought to find a solution to this age-old problem that would transform and prolong the opportunity of safe and secure long-distance sea travel. Such was the problem to mariners that the British Government offered an award of £20,000, around £3million in today's money, for the first person to successfully solve the problem. Harrison made a total of 5 versions of his 'marine chronometer' during several years of design and research.

Harrison's main issues were that of maintaining clock accuracy in a moving ship and how to counter the differences in humidity and temperature in different climates. After many years of frustration, trials and demonstrations to various government committees to prove his worth, he eventually had the backing of parliament and academics alike, having countered many setbacks along the way. He received the first stage payment of £10,000 but had to personally plea with King George III to intervene with the government to get his final award payment. One of the first people to use his maritime clock was Captain Cook on his voyage of discovery. Cook described Harrison's invention as 'our faithful guide through all the vicissitudes of climates'.

Plastic   1855

Alexander Parkes was a British metallurgist and serial inventor from Birmingham, who without help invented fully synthetic plastic. Parkes called his new material 'Parksine' and introduced it as one of the most significant building materials ever invented. He did not get rich from his creation, but out of all the inventions Parkes devised, plastic is now one of the most common building and fashioning materials ever invented.

Plasticine   1899

William Harbutt was a teacher of art, born in North Shields and worked as the headmaster for the Bath School of Arts. In the course of his work at the school, Harbutt developed a new type of modelling clay due to the limitations of natural clay. Critically, Harbutt's new modelling clay was non-toxic and did not dry out like normal clay and in use it was very pliable. He immediately saw the potential of his invention and applied for a patent that was granted to him in 1899 He called this new clay, Plasticine and it went on sale in 1900 as a malleable, more colorful alternative to modelling clay.

Polyethylene   1898

Eric Fawcett & Reginald Gibson were two British industrial chemists working at the ICI (Imperial Chemical Industries) plant in Northwich. Although discovered by accident in 1898 during an experiment, it was never developed or commercialized. In fact, it was not until 1933 that industrial practical polyethylene synthesis was discovered, once again by accident. Polyethylene formed overnight from a leaking test vessel and Fawcett and Gibson were convinced there was an abundance of uses for this substance, prompting them to obtain patents and eventual production. Due to its many top-secret uses during the war years, it was not until the 1950s that polyethylene became common.

Satellites - geostationary   1945

Sir Arthur C Clarke was a British science writer, science fiction author, television host, inventor and undersea explorer. Clarke also had a talent for incredibly accurate predictions of future technology that went from the realms of science fiction to become reality with uncanny accuracy. One such prediction was in his famous proposal of geostationary Satellite communications that was published in 1945 in the Wireless World magazine. Nobody took this seriously at the time, but it actually became reality within 20 years, culminating in the launch of Intelsat 1 Early Bird on April 6th, 1965 becoming the world's first commercial geostationary communication satellite.

Seismograph   **1896**

It was a devastating earthquake that shook Japan, leading British geologist John Milne to invent and develop the seismograph. He relocated to Japan in 1875 as professor of geology and mining, and it was the devastation of the much-documented earthquake in the city of Yokohama during 1880 that spurred Milne to start many years of intensive study into the causes and effect of this great natural disaster. It was during the post-quake period when Milne, born in Liverpool, constructed the world's first purpose-built laboratory to study earthquakes.

After years of thorough research and dedication into pre-quake tremors and aftershock phenomena, Milne unveiled his quake detection machine, called the seismograph, developed with the help of his fellow British scientists, Thomas Grey and James Alfred Ewing. The seismograph detected and measured vibrations in the ground and transcribed them on to photographic paper.

 Silicone   **1899**

Frederick Kipping was a chemical scientist from Manchester who received a PhD from Munich University after studying closed carbon chains. Kipping devoted the majority of his time studying organic silicon compounds and published fifty-four papers on the subject in the period 1899 to 1937. However, he failed to understand the huge commercial potential of his research. The opportunities that Kipping's research had provided were realized by Corning Glass, who with Dow Chemicals formed the Dow-Corning Corporation and in 1943 commenced the manufacture of silicone polymers, now used the world over, primarily as sealants.

 Splitting the atom   **1932**

British physicist John D Cockcroft and Irish physicist Ernest Walton were pioneers in their discovery of how to split the atomic nucleus through the use of a self-built high Voltage device to accelerate the particles to high energy and high speeds. Cockcroft and Walton were influential in nuclear power development.

Science

 ### Spectacles - modern side arm type   1730
British innovator Edward Scarlett invented the world's first modern glasses featuring rigid side arms. Up to this point, early spectacles were held on to the wearer's face by hand or tied with silk or cotton ribbons. Scarlett's breakthrough came in 1730 when he developed rigid side arms that held the spectacles on a wearer's face, resting on the top of the user's ears.

 ### Sunglasses   1752
British inventor James Ayscough started to experiment with tinted lenses in spectacles in the mid 1700s. He was convinced that tinted glasses with a green or blue tint could correct certain sight problems. He was not concerned at the time about protecting people's eyes from the sun; instead it wasn't until 1752 that he found this to be a useful by-product of the tinted lenses, after realizing that his original theory had no vision correcting value whatsoever.

 ### Sunglasses – polarized   1937
Inventor, Edwin Land created polarized sunglasses to protect the eyes from harmful ultra-violet rays. Their design incorporated alternating lenses that shifted the sun's harmful rays in the opposite direction.

 ### Synthetic rubber - isoprene   1891
Sir William Augustus Tilden was a British inventor who produced a new synthetic rubber material from his experiments using the method of destructive distillation of turpentine. Tilden also gave isoprene its structural formula $CH_2=C(CH_3)-CH=CH_2$.

 ### Water desalination   1627
Francis Bacon was a prolific British inventor among other talents. He experimented with seawater desalination, by trying to remove salt particles using a basic sand filter. It was not very effective, but it did lay the foundations for deeper investigation into this ground-breaking research.

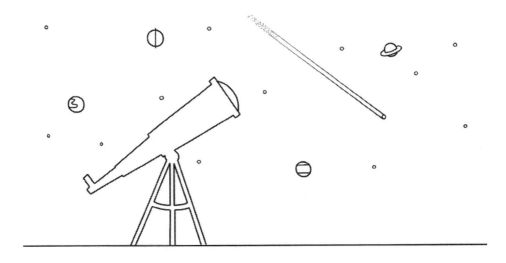

Astronomy

 Binary stars   **1664**

British physicist and astronomer Robert Hooke discovered what is known as the binary star system, Gamma Arietis in the constellation of Aries. The significance of this star system is that it is of a double star type and the orbital period of the two stars is more than 5000 years.

 Black holes   **1974**

Physicist Stephen Hawking stimulated discussions and real recognition in a paper, suggesting that black holes existed in space, even though they appear to break two basic rules of physics; quantum mechanics and Einstein's law of relativity. His paper discussed the 'event Horizon', the area that's considered to exist around a black hole from which nothing can escape. He also proposed that black holes were losing mass and that they would eventually evaporate.

Astronomy

 Communication satellite 1962

Typically, a man-made satellite that is stationed in orbit, in space, for the purpose of telecommunications. Used primarily for fixed point to point services the satellite provides a microwave link using radio relay technology. The communication satellite was invented by NASA aerospace engineer John Robinson Pierce and his first commercial launch was called Telstar, the world's first active communication satellite for telephone and high-speed data.

 Geosynchronous satellite 1963

A satellite whose orbital track on Earth repeats regularly on accurate points over Earth is known as a geosynchronous satellite. The world's first such device was launched on a Delta rocket by NASA in 1963 and was invented by Harold Rosen.

 Halley's comet 1705

Edmond Halley was an astronomer who inspected and researched reports of a comet nearing Earth in 1531, 1607 and 1682. Halley concluded that the three comets were in fact the same comet nearing earth on its return orbit. He predicted that the same comet would appear again in 1758. Halley's calculations proved that some comets do orbit the sun. Although Halley did not live to see the comet's return, his discovery led to the comet being named after him.

 Lunar module 1969

The landing portion of the spacecraft constructed for the Apollo program and built by Grumman, the lunar module was extensively tested on lunar like Earth terrain before one of the greatest moments in history, when man first stepped foot on the moon. A Grumman project engineer, Tom Kelly, is credited as the inventor of the lunar module.

Neptune - discovery   **1847**

John Couch Adams was a scientist from Launceston who, through his passion for mathematics, proved that there must be another planet circling the sun like Earth. He published his conclusion and named the planet Neptune. Adams predicted the existence of Neptune due to irregularities of the orbit of Uranus and decided to investigate if there was another planet behind it.

 Pluto   **1930**

After the discovery of Neptune in 1846, there was much debate surrounding another planet that may be beyond its orbit. Eventually Clyde Tombaugh discovered Pluto in 1930, validating the theory that there was a ninth planet in our solar system.

 Pulsars - discovery   **1967**

During the course of her PhD at Cambridge University, scientist Jocelyn Bell discovered an extraordinary signal at a wavelength of 3.7 metres that she felt was unusual because it was transmitting sharp bursts of radio energy regularly at about one burst a second. This phenomenon was unlike any other signals from known cosmic sources such as solar wind, galaxies or stars. Bell checked her findings with her advisor, radio astrologist Anthony Hewish. They both confirmed that these remarkable findings were synchronized with star time rather than Earth time, suggesting that the signal was definitely extra-terrestrial, not Earth sourced.

Hewish and Bell maintained the tracking of their discovery and eventually, after finding a second unrelated signal pulsing at 1.2 seconds, announced their findings but they were still not certain of the source. This prompted further research in other related areas by respected academics. Ultimately British academic Bell was awarded her PhD and British lecturer Hewish was awarded the Nobel Prize for the discovery of pulsars.

Astronomy

 ### Radio telescope   1932

A radio telescope is quite different from a standard optical telescope as it is effectively a form of directional radio antenna, used in radio astronomy. Operating in the radio frequency sector of the electromagnetic spectrum, they are capable of detecting data and radio sources millions of miles from Earth. Characteristically, radio telescopes are built with a large parabolic dish and antenna used in an array or singularly and was invented back in 1932 by Karl Guthe when he discovered radio static coming from the Milky Way.

 ### Space observatory   1946

A space observatory is any device in outer space, such as a telescope, employed for observation of galaxies, distant planets and stars. Astrophysicist, Lyman Spitzer was the first person to describe the notion of a space telescope, ten years before the Soviet Union launched Sputnik into orbit. Spitzers dream of a space telescope eventually materialised in 1990, when the space shuttle Discovery, launched the Hubble Space Telescope.

 ### Space shuttle   1981

Part of the Space Transportation System (STS), the space shuttle was a reusable spacecraft designed and operated by NASA for human spaceflight operations. The space shuttle was used for servicing satellites, returning apparatus to earth, space observation and later in its capacity as a shuttle to the American-Russian operated space station. George Mueller is credited for stimulating, designing and supervising the program following the closure of the Apollo project in 1972.

 ### Spectroheliograph   1893

Invented in 1893 by Chicago born solar astronomer, George Ellery Hale, the spectroheliograph is an astronomical device that captures a photographic image of the sun at a single monochromatic wavelength of light.

Uranus - discovery 1781

Uranus has of course been seen many times and incorrectly described as a star prior to being correctly identified as a planet of our solar system. However, it took astronomer William Herschel, who was in the back garden of his home in Bath, to incorrectly report it as a comet. Then after being involved in a chain of sightings using a telescope of his own design, he not only compared it to a comet because it had changed its place, but also noted its characteristics and similarity to that of a planet. Later that century, Hershel's observations were confirmed by other astronomers and by 1783, it was established that it was a primary planet of our solar system. Hershel initially named the planet after the King: George's planet. But the name was not popular outside of Britain. Erik Prosperin, a Swedish astronomer, suggested the name Neptune, but the influential astronomer of his time, Johann Elert Bodes choice of name – Uranus, named after his close friend's discovery of Uranium, was widely adopted and accepted by 1850.

 Weather satellite 1959

A type of satellite that is typically used to observe the climate and weather of the Earth. The first weather satellite, Vanguard 2, was created and launched by the US Naval research Laboratory at Cape Canaveral, Florida in 1959.

Biology

 Animal echolocation   **1938**

Also known as biosonar, animal echolocation is the genetic sonar used by many animals such as shrews, bats, whales and dolphins. The term was created by Robert Galambos and Donald Griffin after discovering that it was used by bats.

 Dinosaur - nomenclature   **1856**

Richard Owen was a British paleontologist and biologist who is recognised as being the first person to use the phrase Dinosaur from the Latin word 'Dinosauria' meaning 'terrible reptile', in a paper that he wrote describing fossils.

Evolution - by natural selection   1859

Charles Darwin was a British scientist from Shrewsbury, who created the basis of the theory of evolution through natural selection, transforming existing concepts of the natural world. Prior to Darwin's publication, most people believed that God created the world in seven days. Indeed, there are many that still believe this theory to be true. But on one of his many travels, Darwin had discovered evidence from fossils in rock, suggesting that there were animals that had lived many millions of years ago. On his return to Britain, he tried to discover how species evolved; the outcome of which was his theory of evolution, occurring by the process of natural selection.

 ### Hormones - discovery   1903

Physicists William Bayliss and Ernest Starling introduced the term hormone in a lecture to the Royal College of physicians in 1905. Bayliss and Starling demonstrated that if acid was instilled into the duodenum, then it stimulated secretion in the pancreas. This was the first time that conclusive evidence of this phenomenon had been shown to other scientists.

 ### Rapid eye movement   1952

More commonly known as REM, rapid eye movement is a normal state of sleep categorized as rapid movement of the eyes. There are two classes of REM: tonic and phasic. The phenomenon of REM sleep and its link to dreaming was co-discovered by Nathaniel Kleitman, Eugene Aserinsky and assistant medical student, William C Dement at the University of Chicago.

 ### Red blood cells   1663

British scientist Robert Hooke discovered the scientific theory of cells and the properties of cells. Through his cell research, Hooke was the first person to discover red blood cells, following huge advances in the world of microscope optics and magnification.

Biology

 Stem cells - first practical use   **2008**

British surgeon and physicist Martin Burchall led a team that created a new windpipe, grown by using stem cells, for a 30-year-old Columbian - Claudia Castillo. This was the world's first transplant organ using a patient's own stem cells to grow and replace an organ.

 Test tube baby   **1977**

British physicists Patrick Steptoe and Professor Robert Edwards pioneered the technique of vitro-fertilization. The result of the first successful operation was a baby girl called Louise Brown.

Monoclonal antibodies   **1986**

Antibodies for experimentation and medical use, were previously derived from mice, making them difficult to use on humans due to the human immune system triggering anti-mice responses. It took the inventive approach of British Biochemist, Sir Gregory Winter to pioneer systems to fully humanize antibodies for therapeutic use

> "One general law, leading to the advancement of all organic beings, namely, multiply, vary, let the strongest live and the weakest die."
>
> Charles Darwin

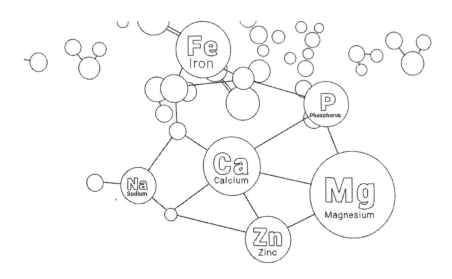

Chemistry

 Argon - discovery and proof   **1894**

John William Strutt and William Ramsay were British physicists who discovered the inert gas, argon, an achievement for which Strutt earned the Nobel Prize for physics in 1904. Argon is used in older style filament lamps to prevent oxidation, thus prolonging its life.

 Babcock - test   **1890**

Stephen Moulton Babcock invented the first successful test used to confirm the fat content of milk. The catalyst for his invention came about after it was discovered that dishonest farmers were diluting full cream milk with water and removing some of the cream, before selling on to factories who were buying the milk in bulk.

Benzene - first isolation　　　　　　　　　　　　　　　　1825

Benzene, a natural hydrocarbon and an element of crude oil, was isolated from this liquid by British scientist Michael Faraday in 1825 for the first time. Faraday was examining a viscous deposit that was a by-product of 'portable gas', made by plunging fish or whale oil into a white-hot furnace. This process created the gas which was condensed and then stored into portable containers for later use, to power gas lamps in public and private residences and commercial premises, such as shop displays. The oily liquid that was produced when this gas was compressed fascinated Faraday.

He successfully reported in a paper to the Royal Society, his successful separation of a new compound of hydrogen and carbon, 'which I may by anticipation distinguish as bi-carburet of hydrogen'. Following Faradays discovery, benzene was being manufactured on an industrial scale by the mid 1800s from coal tar, a by-product of coke manufacturing. It was used mainly as a degreasing solvent, but later in the 1900s was used in all manner of products such as, soaps and aftershave lotions.

 ## Boron - isolation　　　　　　　　　　　　　　　　　　　　1808

Humphrey Davy was a British chemist who isolated boron through the combining of potassium (K) and boric acid (H_3BO_3). It is noted that in the same year and by the same method, French chemists Joseph-Louis Gay-Lussac and Louis-Jaques Thénard also isolated boron.

 ## Chemical electrolysis　　　　　　　　　　　　　　　　　　1807

Sir Humphrey Davy was a British physician and electrochemist who started experimenting with electrolysis by using a battery to prepare elements and split compounds. He discovered new metals such as potassium and sodium and electrolyzed molten salts in 1807, and in the following year he revealed magnesium, barium, strontium and calcium. Davy's research into electrolysis and compound separation led to the new field of electrochemistry being founded.

Chloroform - discovery   1831

A chemical compound that does not combust in air, although it can ignite if combined with flammable substances. Chloroform was discovered by Physician Samuel Guthrie and then some months later in the same year, independently by French chemist Eugene Soubeiran.

 ## Dow process   1891

Herbert Henry Dow was a Canadian born American chemical industrialist and founder of the famous Dow Chemical corporation. The Dow Process is the electrolytic method of the extraction of bromine from salt water or brine and this enabled him to produce bromine on a commercial basis.

 ## Heavy hydrogen   1931

Also known as deuterium, heavy hydrogen is a stable isotope of hydrogen found in huge quantities in the Earth's oceans. Harold Urey is credited with its discovery in 1931.

 ## Heavy water   1933

The isotope deuterium was discovered by Harold Urey in 1931 who later discovered a method of concentrating heavy hydrogen in water, when combined, it is more commonly known as heavy water and a constituent element in the creation of the hydrogen bomb in the 1940's and 1950's.

 ## Helium - first observed on Sun's surface   1868

Sir Joseph Lockyer was a British astronomer from Salcombe who discovered that the Suns atmosphere contained a previously unknown element called helium. It was named after the Greek God Helios, Greek for 'Sun' or 'the Sun God'. Lockyer identified helium, twenty-seven years before it was discovered as an element on earth.

Hydrogen - discovery   1776

Henry Cavendish was a scientist, chemist and natural philosopher. Through experiment, his most notable discovery is that of hydrogen. He explained the concentration of 'inflammable air' that formed water when combusted. Antoine Lavoisier later replicated Cavendish's experiment and named the element hydrogen.

Multiple effect evaporator   1843

As defined in chemical engineering, a multiple effect evaporator is a device for efficiently using heat from steam to evaporate water. The inventor of this apparatus was Norbert Rillieux who devised and patented it in 1843.

Nitrous oxide - discovery   1772

Joseph Priestly, a British clergyman and scientist discovered nitrous oxide, an odorless and colorless gas in 1772. Priestly produced nitrogen oxide by heating ammonium nitrate with iron filings, then siphoning the gas that was emitted, through water to detoxify, thus producing nitrous oxide. Priestley was also famous for isolating other key gases such as; sulphur dioxide, ammonia, carbon dioxide, carbon monoxide and oxygen.

Nylon   1938

Nylon was invented by a team of scientists and researchers working under the leadership of Wallace H Carothers at E. I. Du Pont de Nemours and company in 1938. It was essentially a plastic that could be drawn into strong thin fibers. Famous for nylon hosiery, this very practical invention found use in a whole host of fabric and synthetic rope applications over the years.

Oxygen - discovery   1774

Renowned scientist Joseph Priestly, made the discovery of oxygen 'O' in 1774. He produced oxygen by heating mercuric oxide (Hg0). At the time, Priestly called the gas 'dephlogisticated air', but the name of oxygen was actually created by Antoine Lavoisier who incorrectly thought that oxygen was essential to create all acids. Priestley was also famous for isolating other key gases such as; nitrous oxide, sulphur dioxide, ammonia, carbon dioxide and carbon monoxide.

 ### Periodic table   1865

John Alexander Reina Newlands was a British chemist who developed the Periodic table, now acknowledged as the benchmark in this field the world over. Newlands arranged the periodic table elements according to their atomic weights. He designed the now familiar periods and groups aligned in columns within the periodic table. Reina Newlands also devised the Law of Octaves.

 ### Plutonium   1940

A chemical constituent with the symbol Pu and atomic number of 94, plutonium is a radioactive transuranic synthetic element. Plutonium was co-discovered by J W Kennedy, A C Wahl and Glenn T Seaborg in 1940.

 ### Polyvinylidene chloride   1933

A polymer derived from vinylidene chloride, Polyvinylidene chloride can be found in the production of many household and industrial products. Dow Chemical lab technician Ralph Wiley accidently discovered Polyvinylidene chloride in 1933.

 ### Potassium - first isolation   1807

Humphrey Davy was born in Cornwall. Davy discovered potassium in 1807, extracting it from caustic potash. Prior to the 1800s there was no distinction between potassium and sodium. The first metal to be isolated by electrolysis by Davy was potassium.

Chemistry

 ### PH meter   1935
Invented by Arnold Orville Beckham, the pH meter was a device used to determine the pH level of a liquid, being either acidic or alkaline.

 ### Propane   1910
A three-carbon alkaline, propane is primarily a gas, but can be compressed into a portable liquid. It begins life as a by-product from natural gas processing. Commonly used as fuel for heating, cooking and powering engines, propane was first discovered as a volatile element in gasoline or petrol by Dr. Walter O Snelling, working for the US Bureau of Mines.

 ### Scanning acoustic microscope   1974
A device that uses focused sound to investigate, create an image of an object or to measure, the scanning acoustic microscope (SAM) is commonly employed in failure analysis and non-destructive assessment. The device was invented by two researchers at the Microwave Laboratory at Stanford University: C F Lemons and R A Quate.

 ### Silica gel   1919
A granular, absorbent form of silica made from sodium silicate, silica gel was invented by professor Walter A Patrick at St Johns Hopkins University in Baltimore, Maryland. Silica gel is most commonly used to control moisture levels in packaged perishable goods or sensitive electronic equipment, by absorbing moisture when it exceeds certain levels.

 ### Sodium - first isolation   1807
Humphrey Davy was the first person to isolate Sodium. Davy made this discovery through the electrolysis of NaOH, ultra-dry molten sodium hydroxide gathered at the cathode. Davy went on to isolate potassium using a comparable method the same year.

Thallium - discovery 1861

William Crookes was a physicist and chemist who discovered thallium in residues of sulphuric acid production. Thallium is not found naturally on its own in nature. At the same time as Crookes' discovery, Claude-Auguste Lamy also made the same breakthrough, both using the newly created system of flame spectroscopy.

Viagra 1991

Andrew Bell, David Brown and Nicholas Terrett from Pfizer PLC originally developed Viagra for heart patients suffering from angina, but after proving insignificant in treating heart problems, under the clinical guidance of Ian Osterloh, the team found that there was an unexpected side effect to this drug. Sildenafil as it was then known, could improve and sustain a man's penile erection. Pfizer immediately stopped research on this drug for heart treatment and it is now more widely known as Viagra, the world's first impotence treatment.

 Xerography 1938

Chester Floyd Carlson invented a process that could duplicate copies of documents without the use of ink. To achieve this, Carlson charged an illuminated plate with static electricity, then added a plastic powder to the page to remain white under light and static charge, thus reproducing a duplicate of the original in the remaining areas. The modern photocopier was thus invented and Carlson's Haloid Company was later renamed 'The Xerox corporation'.

> "Living organisms are created by chemistry. We are huge packages of chemicals"
>
> David Christian

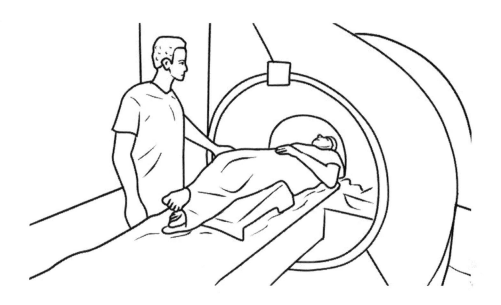

Medical innovation

 ### Artificial heart 1952

Implanted into a donor's body to replace the natural heart, the artificial heart invented by Dr Forest Dewey Dodrill and his first successful operation was performed on 41-year-old Henry Opitek who was a long sufferer of short breath. The artificial heart was used as a temporary measure whilst performing open heart surgery. Although in 1981, Dr Robert Jarvik implanted the world's first permanent artificial heart and kept the recipient, Dr Barney Clark, alive for 112 days.

 ### Breast pump 1854

Invented by O H Needham on June 20th, 1854, the breast pump is a mechanical device that removes breast milk from a lactating female. Today, breast pumps have progressed somewhat, but manual pumps are still available and more advanced electric pumps have become the norm.

Cataracts - artificial lens implant   1949

British ophthalmologist and surgeon Harold Ridley, pioneered cataract surgery. While working on the eyes of a fighter pilot who was the victim of a flying accident, he noted that the Perspex shards embedded in his patient's eyes were not rejected by the body's immune system. This 'Eureka' moment led to Ridley developing the first Perspex intraocular lens to be placed in the eye, enabling millions of people globally to see again, following cataract operations based on his pioneering work.

 ## CAT or CT scanner   1971

Godfrey N Hounsfield was an electrical engineer. His development of the CAT scanner began when he constructed a computer that could process x-rays taken at all angles, thus creating an image of the object in segments. By applying this development to the medical field, he had created a process known as computed topography. Hounsfield then built a trial head scanner and used a preserved human brain and then the fresh brain of a cow for sampling. He next tested the device on his own head. On October 1st, 1971 CT Scanning was adopted by the medical world, primarily for head scans. In 1975, Hounsfield developed and built the first full body CAT scanner.

 ## Cloning - theory   1962

John B Gurdon is a scientist, recognised for his work in genetic cloning and stem cell research published in 1962. Gurdon was the first person to successfully clone an animal from a single cell and revolutionized the long development of stem cell technology into the 21st century.

 ## Cloning - first successful cloning   1996

Dr Ian Wilmutt is a British embryologist from Warwickshire. He is most famous for leading his research team in the first successful cloning of an animal, delivering the now legendary 'Dolly the Sheep'. He was awarded an OBE in 1999 for his services to embryo development.

Colour blindness - discovery 1794
John Dalton accidentally discovered colour blindness when he humiliated his family by buying his mother a set of bright scarlet stockings for her birthday. It was then that he realized that he was blind to this particular colour and his embarrassing experience had assisted him in understanding his own colour blindness because he had thought the stockings were blue. His brother also suffered the same problem. This was when Dalton realized there was a genetic link to colour blindness. His own colour blindness was the most common red-green variation and to this day it is still known as Daltonism.

 ## Compound microscope - 30 X magnification 1665
Robert Hooke was a prolific scientist, architect and inventor and he created the world's first compound microscope. In his publication on the device and its use, he called it the micrographia. Hooke's microscope required the object to be placed on a glass slide and then natural light is diffused through the object, with the final adjustable lens close to the slide on the upper side. Hooke's design made the object appear detailed and clear against the bright background. The same design is still used in school and college laboratories the world over, only now the more sophisticated type uses a light beneath the slide for greater clarity and are commonly known as brightfield or illuminated compound microscopes now.

 ## Combined oral contraceptive pill 1960
An oral birth control pill, the combined oral contraceptive pill is a mix of estrogen and a progestin taken orally. The pill prohibits female fertility and Dr Gregory Pincus's invention was approved for use by the FDA in 1960. Unfortunately, the pill was not commercially cost effective and was in production for only three years.

Contraceptive pill   1968

British chemist and physicist Herchel Smith developed the world's first cost effective and wholly synthetic contraceptive pill. He was born in Plymouth in 1925 and studied at both Oxford and Cambridge Universities. Smith later lectured at Manchester University, where he created and patented processes for making new steroids. He then moved to the USA and created a low-cost method to make a wholly synthetic hormone that prevented contraception called 'norgestrel'. This proved to be the beginning of low-cost birth control. Later, 'Ovral' was developed and became available for general use in 1968. More than 800 patents were held by Smith for his research and production techniques, making him a multi-millionaire. He died in 2001 and left £45 Million to Britain's Cambridge University and several million to Harvard University in the USA.

 ## Controlled clinical trials   1747

James Lind was a naval surgeon who performed what is now recognised as the world's first controlled clinical trial, during which he proved that citrus fruit prevented scurvy, a common disease of the day especially amongst mariners. Through the trials, he discovered that scurvy was caused by a lack of Vitamin C. Scurvy caused more deaths at sea in the 18th century, than enemy action. Lind also discovered that steam condensed from seawater could be drunk safely. It took many years before this process of water distillation was adopted, as there was no realistic method to boil up adequate quantities of salt water until the early 19th century.

 ## Clinical thermometer   1887

Sir Thomas Allbutt was a physician and inventor of the world's first clinical thermometer. The end of the thermometer is put under the patient's tongue, under the armpit or into the rectum to reveal the body temperature of the patient. Allbutt's invention consisted of a thin glass tube with a bulb at one end that contained a liquid that expands uniformly with temperature. The tube has calibration marks along it with mercury or colored alcohol used as the fluid. This method of determining body temperature in a clinical setting was used right up until the 2000s, after which more accurate digital thermometers became available for general clinical use.

Medical innovation

 Defibrillator - first use on a patient   **1947**

A defibrillator is the ultimate treatment for cardiac arrhythmias, ventricular tachycardia and ventricular fibrillation. The device works by supplying a safe dose of electrical energy to the patient's heart, forcing its muscles to spasm and restart its full pumping capacity. The first use on a patient was in 1947 at Case Western Reserve University by Dr Claude Beck. Although, in its modern form, the inventor was electrical engineer William Kouwenhoven in 1930 from the John Hopkins University of Engineering of Baltimore, Maryland.

 Diabetes - first synthetic insulin   **1982**

A team of British scientists from a subsidiary of Eli Lilly invented synthetic insulin. Prior to insulin being discovered, diabetes was a feared and fatal disease. Many doctors realized that sugar worsened diabetes and that a no-sugar diet was the most effective remedy, but not a cure. Then in 1921 a Canadian team of scientists discovered insulin and its effect on the human body. But it was not until the 1980s that a research team of British scientists from the Pharmaceutical Company, Eli Lilly, invented a synthetic substitute to replace bovine insulin along with its many side effects. The team inserted the gene for human insulin into a bacterial DNA and then used the bacteria as tiny factories to produce A and B chains of the protein individually. The second step involved a chemical process to combine them. It resulted in the world's first synthetic human insulin and it was called Humulin. Using Humulin as treatment for diabetes revolutionized not only the industry, but also the lives of millions of diabetic sufferers worldwide, because there are few side effects compared with animal insulin.

 Dissolvable pill   **1884**

Any form of tablet that is ingested orally that is crushable and can be easily digested in the stomach is known as a dissolvable pill, contrasting to the alternative hard coated tablets. The dissolvable pill was invented by William E Upjohn.

Dental chair - fully adjustable   1832

British inventor James Snell form London designed the world's first fully adjustable dental chair. Although the adjustments compared to today's chairs now seem negligible, it shaped future dental chair innovation in the years ahead.

 ### Dental drill   1875

The dental drill is a small high revving tool used in the profession of dentistry, primarily to drill out decay or for reshaping a tooth to accept a false cap or crown. Credit for its invention goes to George F Green of Kalamazoo, Michigan.

 ### Doppler fetal monitor   1958

Also known as a doppler fetal heartbeat monitor, the doppler fetal monitor is a hand held apparatus that uses ultrasound to recognize fetal heartbeat. The doppler fetal monitor was invented by obstetrician Dr Edward H Hon.

 ### Epilepsy - pioneer in treatment   1858

Edward Henry Sieveking was a British physicist, born in London. He discovered that the condition of epilepsy could be treated with the use of bromide and it became the first anti-convulsant in the world. He also published a significant theoretical advancement in epileptology, effectively the end of 'essential epilepsy'.

 ### Gamma camera   1957

Used to image gamma radiation emitting radio isotopes the gamma camera employs a technique known as scintigraphy. The gamma camera is a nuclear medical imaging device enabling the user to view and analyze images and was invented by Hal Anger in 1947.

Glucose meter   1962

Co-invented at the Cincinnati Children's Hospital by Ann Lyons and Leland Clark, the glucose meter is a medical device used to determine the concentration of glucose in blood. The sensor on the meter measures the amount of oxygen consumed by the glucose enzyme.

Hay fever - discovery of   1819

More than one in four people suffer from hay fever in Britain alone and it's hard to believe that just two hundred years ago, nobody knew it even existed. But it took the determination of one man, himself a sufferer, to persuade the medical establishment to truly understand the scale of the phenomenon. John Bostock was a British doctor from Liverpool, whose main expertise was the study of bodily fluids, in particularly urine and bile. At the start of summer every year, Bostock suffered from sinus blockages, a runny nose and was generally lethargic for the next couple of months.

Later in 1819, Bostock decided to write a paper entitled: 'Case of a periodical affection of the eyes and chest'. It was a report effectively about Bostock's experience, detailing the symptoms and attempts of trying to allay the side effects by different treatments. These included, taking a cold bath, self-induced vomiting and taking opium, but nothing appeared to work. Determined to find a cure, Bostock expanded his research looking for and taking data from fellow sufferers.

By 1828 Bostock had located twenty-eight victims of hay fever and it was in a second publication that he named the condition as 'Catarrhus Aestivus' or Summer Catarrh. After this exhausting work, he decided to retreat to the seaside one summer and noted that he was almost rid of this 'summer catarrh'. Yet on his return to the countryside he noted that again his symptoms returned. It was at this point that he realized the link between summer flora and fauna that was in short supply at the seaside, was the most probable cause of the condition. This was confirmed by further research.

Hearing aid - portable   1902

Although original hearing aids had been in use several years prior to this invention in the form of amplified passive trumpets, the portable hearing aid invented by Miller Reese Hutchinson used a body worn amplifier connected to an ear piece, fitting in or behind the wearers ear. This was the first truly portable active hearing aid.

 ### Heart-lung machine   1953

The first successful cardiopulmonary bypass surgery was performed by Dr John Heysham Gibbon. The operation provides artificial blood circulation and oxygenation and additionally, his invention included a pump that by-passed the lungs, which together was known as the heart-lung machine. The machine effectively allowed the surgeon to operate on both the heart and lung whilst maintaining the circulation of blood and oxygen.

 ### Heart-lung and liver transplant   1986

British physicists and surgeons Sir Roy Calne & Professor John Wallwork performed the world's first triple organ transplant on Davina Thompson at Papworth Hospital, Cambridge.

 ### Heimlich maneuver   1974

The Heimlich maneuver is the action of performing abdominal thrusts to specifically rescue choking victims. The rescuer stands behind the victim and then using their hands, applies pressure on the bottom of the diaphragm. The effect of this action compresses the lungs, applying pressure on any lodged matter in the trachea, with any luck expelling it by producing an artificial cough. The inventor of this procedure is Henry Heimlich in June 1974.

Medical innovation

 ### Hip replacement   1961
Sir John Charnley was the pioneer of the world's first hip replacements. British surgeon and physicist John Charley created the world's first hip replacement in 1961 using a femoral stem and ball made of steel, with its hip socket made from Teflon. Several improvements have been made over the years, but more than 100,000 hip replacements are carried out now in Britain alone each year.

 ### Hypodermic syringe   1853
The syringe in its most basic form had been around since ancient times, but it was Alexander Wood's innovative design that for the first time allowed drugs to be administered intravenously without the need to make an incision first in the patient's skin. Legend has it that his inspiration was when a bumblebee stung him. The Hypodermic syringe became a revolution in the world of anesthetics globally.

 ### Iron lung   1928
Invented by Philip Drinker whilst working at Harvard University, the iron lung was a large mechanical device that allowed a person to breathe in the event that normal breathing or muscle control of the lungs had been lost. Drinker had effectively invented the world's first medical ventilator.

 ### Jet injector   1962
A type of medical injecting syringe, the jet injector employs a high-pressure fine jet of injected fluid, rather than use a hypodermic needle to penetrate the patient's epidermis. The jet injector was invented by Aaron Ismach in 1962.

 ### Magnetic resonance imaging   1972
Essentially a medical imaging technique, magnetic resonance imaging or MRI, is most commonly used to visualize the structure and function of the body through high powered magnetic resonance scanning. The inventor of this pioneering device is Dr Raymond Damadian, an Armenian-American scientist. The first successful body scan using MRI was performed in 1977.

NHS - national health service 1948

The NHS was developed and launched on a philosophy that quality healthcare should be accessible to everybody, regardless of wealth. The NHS gave people in Britain, and for the first time anywhere in the world, free Health care that provided on the basis of citizenship rather than ability to pay fees or buy insurance. The person behind this revolutionary way of thinking was British politician Aneurin Bevan who was instrumental in starting this great institution on 5th July 1948. The NHS remains a blueprint for many other countries around the world.

 ### Nicotine patch 1988

A transdermal patch that discharges nicotine into the body, the nicotine patch is used mainly to cut-down or quit smoking completely, although some users remain hooked on the nicotine fix that a patch provides. The co-inventors of the nicotine patch are Daniel Rose, Jed Rose and Murray Jarvik.

 ### Parkinson's disease - discovery 1817

James Parkinson was a British scientist who notably published a paper in 1817 on 'Shaking Palsy', establishing Parkinson's disease as an internationally recognised medical condition.

 ### Peristaltic pump 1881

First patented by Eugene Allen, the peristaltic pump was designed for use exclusively in the transfusion of blood.

 ### PET scanner 1972

A commonly used medical device, the PET scanner examines the whole body to detect diseases such as cancer. The Pet Scanner was co-invented by scientists Michael Phelps and Edward J Hoffman.

Medical innovation

 Defibrillator - portable 1965

Dr James Pantridge was a physician, cardiologist and professor from Hillsborough, who is famous for revolutionizing emergency medical aid and improving paramedic services, through his development of the portable defibrillator. This machine treats life threatening cardiac attacks, by delivering a pulse of current to the heart, controlled by the operator. This invention is responsible for saving the lives of many thousands of people world-wide due to its portability and local availability, without the need to visit hospital.

 Teflon 1938

Teflon is a synthetic fluoropolymer, better known in the field of chemistry as polytetrafluoroethylene (PTFE). Most associated now with the Dupont brand 'Teflon', it was discovered like so many inventions accidentally by an employer of Kinetic Chemicals, known as Roy Plunkett in 1938.

Medicine

 Acetylcholine 1914

Henry Hallet Dale was an inventor who discovered acetylcholine, the chemical compound that is a neurotransmitter. It was noted for its actions on heart tissue in particular and was confirmed as a neurotransmitter later by Otto Loewi. Both Scientists received the 1936 Nobel Peace Prize in physiology or medicine for their pioneering work in this field. Acetylcholine functions both in the central nervous system (CNS) and in the peripheral nervous system (PNS).

 Anesthetics - nitrous oxide 1799

Nitrous oxide was first discovered by Sir Humphrey Davy and is recognised as the world's first commercial and practical anesthetic, now better known as laughing gas, still used to this day worldwide in many hospitals as a reliable, safe form of anesthesia with very few side effects.

Anesthetics - chloroform — 1847

British obstetrician Sir James Young Simpson was a hugely important figure in the chronological history of medicine. He discovered the anesthetic properties found in chloroform and introduced it successfully for general medical practice during an experiment with some of his friends, where he soon discovered that it could be used to put someone to sleep. It was later discovered that chloroform was highly toxic to the liver.

Anesthetics - halothane — 1951

Discovered by chemist, Charles Suckling while working at ICI (Imperial Chemical Industries). He first synthesized halothane, then an extremely volatile anesthetic in 1951, whilst working in his ICI laboratory in Widnes Lancashire. Following Suckling's first investigations on the anesthetic properties of halothane by experimenting on houseflies and mealworms, he then passed it on to a renowned pharmacologist, Jaume Raventos for further evaluation on other animals.

After establishing its medical properties and possible application, Reventos then provided his findings to a prominent Manchester anesthetist, Michael Johnstone, who recognised its huge advantages over conventional anesthetics, establishing its first clinical trial in 1956. This has been documented and recognised as one of the first instances of modern drug design in the world.

Antiseptic surgery — 1865

Joseph Lister was a surgeon who introduced antiseptics to surgery and subsequently employed them to increase the success rates in theatre. Open wounds offered easy access for germs to enter the body. Lister read that carbolic acid had been used to treat raw sewage and to kill cattle parasites. This provided him with the notion that it could also prevent infection of wounds. He then started to use carbolic acid to clean the wounds of his patients and at the same time soak the dressings in the same antiseptic fluid. Over time, Listers death rate in surgery decreased from 45% to below 15% after introducing his new antiseptic treatment.

Aspirin - discovery of the active ingredient　　　　**1763**

Edward Stone was a Church of England clergyman who was walking through a meadow one day near his home at Chipping Norton, suffering from aches and pains. He was alleged to randomly rest by a willow tree and broke off a small piece of its bark and started nibbling at it. He noted its remarkably bitter taste and knowing that the bark of the Peruvian tree cinchona produced quinine with a similar bitter taste (used to treat malaria) he deduced that willow might also have similar therapeutic qualities.

Following several experiments of drying willow bark and pounding it to a fine powder, he gave it to about 50 different people to test.

Stone concluded that it was found to be very effective in curing fevers and relatively mild disorders such as headaches. He discovered in the process of testing that the active ingredient was salicylic acid. He sent a letter confirming his discovery to the president of the Royal Society. The letter survives to this day.

 Aspirin - effectiveness　　　　**1982**

John Vane was a British pharmacologist who was a key figure in understanding how pain and anti-inflammatory effects were produced by the drug Aspirin. His research led to new treatments for blood vessel and heart disease, including introducing ACE inhibitors. Vane was co-awarded the Nobel Prize in physiology and medicine in 1982 with Bengt I. Samuelsson and Sune K. Bergstrom for their findings relating to prostaglandins and biologically active substances.

Beta blockers - propranolol   **1964**

The science of cardiology takes in some of the most respected names in medicine. One such person was physicist Sir James Black who contributed hugely in this specialist field, as a scientist and a physician. His work in cardiology and invention in 1964 of Propranolol was a beta-adrenergic receptor antagonist and it revolutionized the treatment and stability of the heart condition angina pectoris. It is regarded as one of the most significant contributions to pharmacology and clinical medicine of the 20th Century. In 1988 he was awarded the Nobel Prize in medicine.

 Blood circulation - discovery   **1628**

William Harvey was a physician and the first person to demonstrate and prove that blood circulates around the body. Previous to this discovery, it was commonly thought that blood flowed freely throughout the body similar to a bag full of liquid, rather than being transported via a series of arteries and veins.

 Blood pressure - measurement   **1733**

British chemist and botanist Stephen Hales, famously took blood pressure measurements in several species of animal by inserting very fine tubes into veins and arteries, then measuring the height that the column of blood rose. This was to be the world's first practical blood pressure test.

 Blood transfusions - human to human   **1818**

British physician, innovator and pioneer James Blundell performed the world's first human blood transfusion. It was in the pursuit of safe and effective blood transfusion that Blundell was most interested and motivated. He first experimented with whole blood transfusion between dogs before successfully carrying out experiments from human to human using syringes, taking care to avoid air intake into the veins. Blundell also started to perfect a device that now resembles a modern-day drip to introduce appropriate drugs in a 'controlled manner'.

Blood transfusion service 1921
British physician Dr Percy Oliver started the world's first voluntary blood transfusion service. Under Oliver's supervision, British Red Cross members elected to give blood at King's College Hospital London; the start of the world's first Voluntary Blood transfusion service.

 ### Brain 'locator GPS' - discovery 2014
A British based research team led by Professor John O'Keefe with May-Britt Moser and Evard Moser shared the Nobel Prize for physiology or medicine in 2014 for discovering and explaining the brain's 'GPS' system. The team based at University College London, discovered how the human brain knows where we are and how to plot a course from one point to another. It is hoped that this discovery will help solve some of the unanswered questions relating to Alzheimer's disease, where patients cannot recognize their surroundings. Professor O'Keefe discovered the first part of the brains positioning system as far back as 1971. Then in 2005, husband and wife team May-Britt and Edvard Moser found a different area of the brain that performs similar to a maritime chart with grid like sections. This they found helps humans to navigate and judge distance. Moreover, according to Dr Colin Lever from the University of Durham (one of Professor O'Keefe's team), he found that the pioneering work and discovery, supports the theory that humans possess an autobiographical memory.

 ### Cholera - discovery that it was water bound 1854
John Snow was a physician, expert in medical hygiene and anesthesia. He traced the source of a major cholera outbreak in London in 1854. His findings pointed to the water borne diseases carried within London's poor water and waste systems and this led to similar recommendations for change in other large British cities. His research ultimately led to important improvements to general public health worldwide.

DNA - discovery 1953

Francis Crick (GB), James Watson (USA), Maurice Wilkins (NZ) won the Nobel peace prize for the discovery of the structure of DNA in 1962, although they first made the discovery and published their findings on the structure of Deoxyribonucleic Acid (DNA) in April 1953. Rosalind Franklin (GB) was also working with Wilkins at King's College London and were both using the technique of X-Ray diffraction to study the DNA. Shortly after, they used their findings to publish their own work on their discovery, a molecular structure of DNA based on all previous research know characteristics, and the now famous double helix. They used this model to explain how DNA replicates and its relationship to hereditary information coded on it.

Female Doctor 1857

British Physician, doctor Elizabeth Blackwell was the world's first woman to receive a medical degree in the USA and become the first woman to be registered on the British Medical Register. Blackwell later broke new ground in encouraging the education of women in medicine, both in the USA and Great Britain.

Gray's anatomy 1858

British anatomist Henry Gray studied the development of the spleen and endocrine glands in 1855. It was through this research that he made a proposal to his colleague Henry Vandyke Carter, with an idea to develop and publish an accessible textbook that would be suitable for medical scholars. Vandyke Carter was agreeable to this project and so the two set out on their crusade of dissecting unclaimed bodies from mortuaries and workhouses for the next two years, to gather data and drawings that formed the basis for their book. The book was completed and published in 1958 in Britain and the following year in the USA. It was called Gray's Anatomy. It was the first published anatomical textbook in the world.

Hepatitis B virus 1964

While working at the National Institute for Health, Baruch Blumberg discovered the hepatitis B virus in 1964.

 ### Hepatitis B vaccine 1976

Twelve years after discovering the hepatitis B virus, Baruch Blumberg finally, through huge endeavor, developed a diagnostic test and successful vaccine for the hepatitis B virus in 1976.

 ### Hypnosis 1841

James Braid is recognised as the first authentic hypnotherapist. Braid was a scientist and practicing surgeon who was the single most significant innovator in the field of hypnosis. Braid saw the benefit and treatment by hypnosis as complementary to primary medicine, rather than an alternative. In his experimental stages, Braid revealed that hypnotism on a human specimen simply set off a chain of invisible sensual events that merely required gentle support to achieve their aims and progress. The modern science of hypnosis certainly owes its name and functional principal to Braid's extensive research and development in this field.

 ### Ibuprofen 1961

Ibuprofen is a drug now used the world over and belongs to the drug family known as anti-inflammatories and the world's most popular drug in that group. Chemist Stewart Adams and his team, working for drug Company Boots PLC in Nottingham, produced this ground-breaking drug after several years of research and patented it in 1961. Ibuprofen is now one of the most widely used drugs in the world.

IVF 1978

Patrick Steptoe was a British pioneer of fertility treatment, gynecologist and obstetrician & Robert Edwards was a biologist and physiologist. Together they broke new ground in the field of in vitro Fertilization. The first test tube baby in the world was Louise Joy Brown and she was born on July 25th, 1978. Steptoe died in 1988, but Edwards was awarded the Nobel Prize in physiology or medicine in 2010. Unfortunately, Steptoe was not considered for the Nobel Prize because it is not awarded posthumously.

Niacin 1937

Also known as Vitamin B3, niacin is water soluble and prevents the pellagra disease. Niacin was discovered while investigating the properties of livers, by Conrad Elvehjem who later discovered its active ingredients in 1937.

Malaria 1897

Ronald Ross was a British Army Surgeon who established the cause of Malaria by discovering the malarial parasite in the infected gastrointestinal tract of mosquitoes. Ross's documented discovery paved the way for other scientists to find a cure to combat the fatal disease.

MRI - Magnetic Resonance Imaging 1973

Sir Peter Mansfield, a physicist from London, is credited as demonstrating how radio signals from MRI can be mathematically evaluated. Mansfield's research enabled a system to be devised in 1973 that could extrapolate the signals of an MRI device and transpose them into a practical image. Mansfield's work is also acknowledged to have made rapid practical Magnetic Resonance Imaging possible by developing the 'MRI Protocol' known as Echo Planar Imaging.

MRI - machine   1980

The world's first full body MRI scanner was built and trialed by a team led by professor John Mallard including professor James Hutchinson and doctor Bill Edelstein. On August 28th, 1980 Mallard's Magnetic Resonance machine based on Sir Peter Mansfield's MRI Protocol, obtained the first clinically useful image of a patient's internal organs. The first patient it was used on following trials was a man from Fraserburgh revealing a cancerous stomach tumor. The MRI scan revealed that the patient also had secondary cancer in his bones and an abnormal liver.

 ### Penicillin   1928

British physicist and inventor Sir Alexander Fleming discovered how penicillium notaturn reacted on bacteria and made the forerunner to modern antibiotics, as we know them today.

 ### Penicillin - mass production   1945

British biochemist Dorothy Crowfoot-Hodgkin discovered the chemical structure of penicillin and through her pioneering research work, it is now the world's most widely used anti-biotic. Later that same year, a team of scientists at the world-famous Oxford University, developed a way to mass-produce the drug. The ground-breaking team included British scientist Norman Heatley, Australian scientist Howard Florey and German scientist Ernst Boris Chain. Chain and Florey shared the 1945 Nobel Prize in Medicine with Sir Alexander Fleming for their work.

 ### Polio vaccine   1952

Vaccination works by priming and stimulating the immune system for a response through an infectious agent. The polio vaccine immunity development effectively prevents human to human transmission of wild poliovirus, protecting vaccine recipients. It was discovered and developed by Dr. Jonas Salk in 1952. He later published his findings in 1954 through the American Medical Association.

Medicine

 ### Smoking - lung cancer link   1950
British physicists Sir Richard Doll and Sir Austin Bradford Hill first proved that smoking causes lung cancer. At the time of their discovery, more than eighty percent of the British population smoked cigarettes, causing a shocking rise in the number of smoking related deaths from Lung Cancer. Bradford-Hill and Doll were asked to investigate this alarming phenomenon. At first it was thought that the rise in lung cancer directly correlated with the huge road-building programme, the fumes from Tarmac-laying and increased car fumes, as ownership was rapidly increasing. But then the breakthrough came; in a sample of seven hundred lung cancer patients, only two were non-smokers. The study of lung cancer patients continued, and it provided conclusive evidence that smoking caused the disease.

 ### Steri-spray - sterilization of water   2008
Ian Helmore was a plumber whose job it was to sterilize water tanks to prevent the breeding of legionella. He discovered that bacteria might remain in the last few inches of pipe work, so while experimenting to find a solution, he discovered that installing an Ultra-Violet lamp inside a spray head or tap would deal with the issue. The Steri-spray instantly sterilizes water. This simple system is now used in the NHS and hospitals the world over.

 ### Surgical forceps   1727
British scientist Stephen Hales is attributed to the invention of surgical forceps. It was and still is a hand-held device for holding or grasping objects, often used when fingers or hands would be too large to hold small items.

 ### Typhoid - discovery and prevention   1838
William Budd was a medical doctor whose investigative work on typhoid fever had huge significance in the years to come. Budd served his early years as a trainee doctor with French surgeon and pathologist Pierre Louis, who had a special interest in an illness of the gut called putrid fever. Budd

was also influenced by the work of another French doctor, Bretonneau who also worked on treating a similar fever. It was Budd's experience in France that encouraged him to look at a similar disease in Britain known as Typhoid Fever, due to similarities to typhus, though not connected.

Budd ultimately discovered the casual bacterium, salmonella typhi. He first published a paper on the subject in 1938 but made his discovery proper 42 years later in 1880. He concluded that rather than being a contagious disease, typhoid fever was in fact caused by poor sanitation and water hygiene and the subsequent ingestion of contaminated water and food. This led him to become an enthusiast in disinfection and the promotion of clean water and drainage systems.

Typhoid - first vaccine   1896

British Medical doctor, Almroth Edward Wright was a supporter of William Budd's discovery that typhoid fever is non-contagious and its causes thereof. Wright set about experimenting with an anti-dote to the fever and in the course of his research, developed a successful inactivated whole-cell typhoid vaccine launched in 1896.

However, the vaccine was found to have some serious side effects and its use was stopped a few years later. To date, although several vaccines are on trial, not one has been sanctioned for commercial use since Wright's invention, with doctors having to use strong antibiotics, rather than administer immunization without side effects, still the long-term goal.

Ultrasound imaging   1956

British physicist Ian Donald was the first person to introduce ultrasound into medicine for diagnostics in 1956. With the aid of a one-dimensional amplitude mode (A Mode) device, he was able to measure the diameter of the fetal head. Then, two years later, Donald displayed an ultrasound image of a female genital tumor. Ultrasound scanning is safe and used throughout the world now to scan and display various stages of the unborn child.

 Vaccination 1796

Edward Jenner was a medical doctor from Berkeley, Gloucestershire, who pioneered the small pox vaccination, and was known as 'the father of immunology'. Jenner noticed that milkmaids who suffered from the mild disease of cowpox, never ever contracted smallpox, one of the biggest killers of the time. So, he famously placed pus taken from a cowpox cyst, into an incision on eight-year old James Phipps arm. Jenner was able to prove that Phipps was immune to smallpox, having been vaccinated with cowpox. He later carried out several more vaccinations on other young children including his own son, to prove his theory. The word vaccine was taken from the Latin word 'vacca' for cow.

 Vitamin A 1917

A bi-polar molecule that is formed with bi-polar covalent links and hydrogen, Vitamin A is linked to a family of similar forms of molecules known as retinoids, that complete the remaining vitamin sequence. Vitamin A can be found in many forms such as animal-based foods. Vitamin A in its common form was co-discovered in 1917 by Layfayette Mendel and Thomas Osborne from Yale University.

 Vitamin E 1936

Chemical compounds that comprise Vitamin E activity, are classed as tocopherols, a series of organic compounds comprising methylated phenols. During experiments with rat feeding, Herbert McLean Evans established that aside from Vitamins B & C, an unidentified vitamin existed. Although all other nutrients were present, the rats remained infertile. It was not until 1936 that Herbert McLean and his colleague K S Bishop discovered the missing substance, named Vitamin E.

> "The art of medicine consists in amusing the patient while nature cures the disease"
>
> Voltaire

Physics

 Ant robotics   1991

Swarm robots are basic low-cost robots with minimal computing and sensing capabilities, known generally as ant robotics due to their behaviour, making them feasible to deploy as teams detecting objects in their path and relaying the information back to a central controller. Their inventor, MIT employee and electrical engineer James Mclurkin, was the first person to conceptualize the idea.

 Argon laser   1964

Invented in 1964 by William Bridges, the argon laser is just one of a series of lasers that use a noble gas as the active medium.

 Atomic clock   1949

Using an atomic resonance frequency standard for its timekeeping component, the atomic clock was invented and developed in 1949 and is credited to the United States National Bureau of Standards.

Physics

 Audiometer 1922

Designed to evaluate hearing loss, an audiometer comprises embedded hardware linked to a pair of headphones and a response button. When a frequency is heard by the patient, they must immediately press the button that is registered by the machine. The audiometer was invented by Dr Harvey Fletcher of Brigham Young University and this method of testing hearing is still used in hospitals and clinics the world over.

 **Automatic volume control** 1925

Generally, a variable system discovered in many electronic apparatus, the automatic volume control or AVC, was invented by Harold Alden Wheeler. The AVC uses a method of control that measures the average output signal level, that is then fed back to automatically adjust the gain to a pre-determined level selected from a range of output settings, normally 1 to 10. This method of volume control is still in use today.

 **Bolometer** 1878

Invented and developed by astronomer Samuel Pierpoint Langley, a bolometer measures the energy of incident in electromagnetic radiation.

Bragg's law 1912

British scientist and physicist, William Lawrence Bragg submitted his derivation of the reflection condition to the Cambridge Philosophical Society at a meeting in November 1912. His paper was published in 1913. The Bragg's law concept provides the condition for a plane wave to be diffracted by a collection of lattice planes. Its use in the development of electro-magnetic radiation, particle waves, gamma rays and x-rays in relation to electrons and neutrons has been invaluable to modern day physics and its progress.

Chemical laser 1965

Obtaining its energy from a chemical reaction, a chemical laser can achieve a continuous wave output within an integrated circuit, capable of attaining mega Watt levels. Used for drilling and cutting in commercial and industrial applications where precision is paramount, the chemical laser was co-invented by George C Pimentel and Jerome V V Kasper in 1965.

 ## Circuit breaker 1836

Designed to automatically operate an electrical switch that will protect a circuit from damage caused by short circuit or overload, a circuit breaker is a pioneering electrical circuit device, still used globally to this day in all manner of applications. The circuit breaker was invented by Charles Grafton Page.

 ## Conservation of energy - law 1843

James Joules was a brewer, who at the age of 22 revealed that heat generated in a coil of wire is directly proportional to the square of the current running through it. The effect is now called 'Joule Heating' and in recognition of Joules' importance in demonstrating energy conservation, the SI unit and the scientific term known as the 'Joule' is named after him.

 ## Cosmic radio waves 1931

A subfield of astronomy, radio astronomy studies celestial objects using radio frequencies. Whilst tracking electrical interference on telephone transmissions, Bell Telephone Laboratories engineer Karl Guthe Jansky, discovered radio waves originating from stars outside our galaxy. Hence, the exciting new field of radio astronomy was formed.

Physics

 Crystal oscillator 1918

An electronic circuit that employs the mechanical resonance of a vibrating piezoelectric crystal material to create a signal with an exacting frequency is called a crystal oscillator. Most commonly used in recent years to maintain accurate timing and used in devices such as quartz watches and clocks, the crystal oscillator was originally invented and developed by Alexander M Nicholson.

 Cyclotron 1929

Invented by Ernst O Lawrence of the University of California, the cyclotron is a particle accelerator that speeds charged particles by applying a high frequency alternating voltage.

 Dalton's law - of partial pressures 1801

Eminent scientist John Dalton discovered the relationship of the individual pressures of each gas in a mixture of gas. Dalton's Law states that the combined pressure of a mixture of gases is equal to the partial pressures of the component gases.

 Diamagnetism 1845

Prolific scientist and inventor Michael Faraday discovered that several substances show a faint repulsion when introduced to a magnetic field and he named this phenomenon diamagnetism. In 1885, Faraday also observed that the he could affect a ray of light in the direction that it was moving by introducing a magnetic field into its path. He noted that he had 'succeeded in illuminating a magnetic curve'.

 Dimmer switch 1892

A dimmer switch is a device that is employed to adjust the light intensity of a light source such as a common lamp or light bulb. They work by varying the voltage and corresponding power to the lamp that controls the output of light. The dimmer, now used in movies, on stage, in restaurants, homes and offices, was invented by Granville Woods. He was the first American of African descent to become an electrical-mechanical engineer and he filed more than 60 patents during his professional career.

 Earth inductor compass   **1924**

Invented and developed by an employee of the Pioneer Instrument Company, Morris Titterington. The principle of the earth inductor compass is that it uses electromagnetic induction by using the earth's magnetic field as the induction field for an electric generator. By the variation of voltage generated, this allows the compass to accurately locate position and direction.

 Electric transformer   **1821**
Electric generator   **1831**
Electric motor   **1831**

There is good reason why all three pioneering inventions above are grouped together. Michael Faraday is best known for his discoveries of the laws of electrolysis and electromagnetic induction. Faraday's greatest breakthrough was his invention of the electric motor in 1831.

Faraday's first development was the electric transformer in 1821. Then ten years later, he created an ingenious device based on a ring and discovered when an iron core was rotated in the centre of the ring of wound wire; it generated a potential across the two wires. This was the world's first generator. Further experiments followed on the same device, when he applied a potential or Voltage to the two wires on the generator, then the iron core would rotate. This was the world's first electric motor. All three related inventions were absolutely pivotal in industrial progress and are probably the three most important inventions of the 19th Century.

Faraday went on to develop many other significant discoveries, such as benzene, the 'Faraday effect' and the legendary 'Faraday cage' and finally he had the unit of electrical capacitance; the 'Farad', named after him. Faraday refused the title of a Knighthood and presidency of the Royal Society, instead devoting his working life entirely to scientific research.

Electro magnet — 1823

William Sturgeon was an inventor and self-taught physicist and mathematician, who in 1823 demonstrated his first simple electromagnet. He revealed its power by lifting a nine-pound weight with just a seven-ounce rod of iron, wrapped with eighteen turns of bare wire that was energized with current from a single cell.

Electron — 1897

Although many scientists before 1897 had suggested that the atom was constructed from a more elementary unit, it was British physicist and inventor JJ Thomson, who was the first to propose that the elementary unit was more than one thousand times smaller than an atom. Thomson advocated that there was a subatomic particle, now known as the electron.

Faraday cage — 1836

Michael Faraday was a prolific scientist, electrical engineer and inventor, who developed a shield made of conducting material, normally mesh, to block external electrical fields, both static and non-static, therefore providing a constant Voltage on all sides of the enclosure. The Faraday cage is used in situations where electronic equipment needs to be protected from static or non-static discharges.

Fluxgate magnetometer — 1940

Measuring the direction and magnitude of magnetic fields, a fluxgate magnetometer uses sensors strategically positioned to assess precision data and evaluation. The device was invented by Victor Vacuier in 1940 whilst working in Pittsburgh for Gulf Research.

 **Fuel cell** 1842

Physical scientist and judge William R. Grove discussed the universal theory of the conservation of energy, but more importantly, pioneered the development of fuel cell technology. He developed a device called the 'gas Voltaic battery', that was to become the world's first working fuel cell that produced electricity. Grove's device produced a current by mixing oxygen and hydrogen. He developed and demonstrated, that the cell could prove that steam can be converted into hydrogen and oxygen and the same process could then be reversed. Grove was the first person to show thermal disassociation of molecules into their component atoms.

 Fusor 1959

In contrast to most controlled fusion systems that slowly heat a magnetically contained plasma, the fusor forces high temperature ions straight into a reaction vessel, avoiding a conventional fusor's complexity. The fusor was invented and developed by Philo T Farnsworth to create nuclear fusion.

 **Graphene** 2004

Scientists recognised the existence of graphene and its remarkable physical characteristics and potential applications for the future. But, the big breakthrough came when two British based researchers at the University of Manchester, Professor Andre Geim and Professor Kostya Novoselov, isolated graphene in 2004. Even though physicists were aware that one atom thick, 2D crystal graphene existed, up to this point, nobody had worked out how to extract it from Graphite. Forecast to transform the world we live in, it is completely transparent, yet conducts both heat and electricity and is billed as the world's first 2D material. Graphene is a technology that has the potential to disrupt medicine, travel electric vehicle technology and a host of other industries. Both Scientists received the Nobel prize for Physics for their ground-breaking work.

Physics

✹ Gravity - the law of   1665

Legend has it, that physicist Isaac Newton was one day sitting under an apple tree, when an apple fell on to his head. It was at this moment that he started to think about the law of gravity. Although this is almost definitely not the case, the story acts as a perfect anecdote to describe what actually happened. What is sure is that Newton started looking into why things fall or travel and the cause and effect of such a phenomenon.

Newton deduced that if a matter is accelerated through falling from a tree, then because its speed changes from zero on the tree, it moves towards the ground, therefore there must be a force in action and this force must be due to gravity. Newton reached the conclusion that any two objects in the universe will exert gravitational attraction towards each other and that the force will have a common form.

Hydroelectricity   1880

William Armstrong was an inventor and engineer from Newcastle and is credited with designing and installing the world's first house in the world, to be powered by hydro-electricity. Hydro-electricity is now the most widely used form of power in the world.

Induction motor   1888

Famous for his invention and development in AC (alternating current), Nikola Tesla, a Serbian-American inventor developed a new method of driving an electric motor using AC induction. Prior to this breakthrough, all motors had brushes that came into contact with the spinning coil shaft, thus creating sparks in its early development, causing wear and tear on the brushes that needed regular replacement. Tesla's induction motor produced torque through electromagnetic induction from the electro-magnetic field of the stator winding, rather than using mechanical contact.

Integrated circuit 1958

An integrated circuit is now ubiquitous throughout the world of electronic design and application, but in the original invention by Jack Kilby in 1958, he successfully integrated large numbers of micro-transistors into a small chip, all created with a piece of germanium that he tested with an oscilloscope. He noted that the oscilloscope showed a continuous sine wave, proving that his integrated circuit was a success. He successfully patented his 'solid circuit made of germanium' later in 1958.

 Isotopes - first concept 1912

Economist and chemist Frederick Soddy proposed that the same elements exist in different forms. He cited that nuclei having the same number of protons could have different numbers of neutrons. So, his theory of isotopes explained that different elements can be indistinguishable chemically, but at the same time they can have completely different atomic characteristics and weights. His theory led to a huge breakthrough in 1920 when he displayed the significance of isotopes, in calculating geological age. This led to the huge breakthrough of carbon dating.

 Jet propulsion - theory 1686

British scientist Sir Isaac Newton developed the principal of jet propulsion. He also forecast that one day people would be able to travel at speeds up to '50mph' in a jet powered vehicle, based on his 'third law of motion' theory 'to every action there is an equal reaction'. But it wasn't until 1774 that a man called Gravesande developed a steam powered Jet Car, when a boiler mounted on the cars chassis pushed the vehicle along by emitting a powerful jet of steam.

Physics

 ### Joule - thermodynamics　　　　　　　　　　　　　　1843

British scientist James Prescott Joule discovered the relationship between heat and mechanical movement. Joule's discovery then led to the law of conservation of energy, in turn leading to the first law of thermodynamics. Joule also worked with Lord Kelvin and helped develop the absolute scale of temperature. Joule further observed the relationship between the current flowing through a resistor and the dissipated heat. This is now called Joules first law.

 ### Laser　　　　　　　　　　　　　　　　　　　　　　1957

A device that emits electromagnetic radiation via a process known as stimulated emission, a laser is usually spatially consistent; a phenomenon that produces light in a very narrow beam but can be converted into a wider spread of light with mirrors and lenses. Lasers are used in many products such as bar code readers, compact disc players, guided missiles, to remove ulcers, cutting steel and can be used to precisely measure range such as the distance between Earth and the moon. The laser was originally invented by Physicist Gordon Gould, who first created the theory in principal and what its uses may be in 1957.

 ### LED - theory　　　　　　　　　　　　　　　　　　1907

The first report of a solid-state light emitting diode being made and tested, was in 1907 by British engineer H J Round of English company Marconi Labs. He was a personal assistant to radio pioneer Guglielmo Marconi. Round used a crystal of silicon carbide and a cat's whisker detector. He was educated for much of his schooling at Cheltenham Grammar School for boys, and later attended the Royal College of Science, where he was awarded a first-class honor's degree. He was successful in submitting and receiving 117 patents.

LED - light emitting diode - commerciality   1962

The world's first practical visible-spectrum light emitting diode was invented and developed by Nick Holonyak Jr. Based on the original work carried out in Britain's Marconi labs in 1907 by British engineer H J Round, Holonyak improved on the original invention and design by making LED's commercially viable both in production and application.

Light bulb - electric incandescent   1800

Eminent scientist Sir Humphrey Davy demonstrated the world's first incandescent light bulb in 1800, by passing electricity through a strip of carbon. The carbon glowed brightly, producing an arc light, but soon burnt out after a few seconds. Davy later experimented with platinum and used a vacuum pump to extract air from the lamp, before it was sealed to ensure the filament lasted longer.

 Light bulb   1880

Joseph Swan created the world's first practical light bulb, beating Thomas Edison to the patent by producing a reliable alternative to the gas lamp. But for both men, their creations had extremely short life spans of just a few hours. Eventually, both men decided to pool their resources and co-develop their inventions, by forming the Anglo-American - Edison Swan Company.

Light switch - quick break technology   1884

British inventor John Henry Holmes from Newcastle-upon-Tyne invented the world's first quick-break light switch in 1884. Holmes solved the problem of switches arcing each time the circuit was opened or closed and solved the solution, by ensuring that the contacts opened or closed rapidly each time the switch was operated. Less arcing meant longer life contacts and the Holmes switching principle is still used internationally in all light switches.

Physics

 Light - nature of electromagnetic theory **1873**

Professor and scientist James Clerk Maxwell, experimented with magnetism and electricity, concentrating on electromagnetic theory and light, discovering in the process that electricity travels at the speed of light. He also recognised that magnetism and electricity are one of the same thing, known as electromagnetism, and in the process of his research predicted the existence of radio waves.

 Lightning rod **1749**

A Lightning rod is a key component of a lightning protection system. The rod is usually linked by thick copper straps from the rod, to the highest point on a building, via a spiked lightning conductor or finial, mounted higher than the tallest part of the building structure. Benjamin Franklin decided to prove his theory that a lightning arrestor system would shield people and buildings from lightning strikes, proving that electricity and lightning were identical and one of the same thing. This principle of lightning protection is now used throughout the world to protect tall buildings or structures.

 LCD Liquid Crystal Display **1970**

It is quite possible that you are reading this article on a device, made possible only because of British inventor, George Gray's creation of the LCD. Gray developed the molecules that enabled LCD's to function. In 1970, Grays work on LCD's was first published, initiating a multibillion-pound industry, making today's profusion of LCD flat screen gadgets achievable.

 Light switch – toggle **1916**

The most common of all light switches is the toggle switch, co-invented by Morris Goldberg and William J Newton of Lynbrook, New York. The key to this design's success is that the toggle handle does not control the switch directly. Instead, it uses an arrangement of levers and springs inside the switch to produce a stable and safe switching device with no noticeable arc or electrical crack under load.

Lithium - ion battery　　　　　　　　　　　　　　**1976**

The world's most popular rechargeable battery started life with a team of researchers and scientists at Oxford University's inorganic chemistry laboratory. In the mid-seventies, there was an upsurge in development by companies and scientists, searching for more compact, lighter and longer lasting power storage solutions, that had hitherto been thwarted by a succession of setbacks.

Led by John Goodenough, his research group included, Dr Phil Wiseman, Dr Koichi Mizushima and Dr Phil Jones, the team looked at various ways of employing lithium cobalt oxide. After much experimentation (and several fires in the lab), the group published in the Materials and Research Bulletin in 1980. The first commercially manufactured Lithium-ion battery was subsequently launched by Sony 10 years later.

The lithium-ion battery is now ubiquitous and is powering the huge growth of electric transportation globally, while the world seeks to reduce carbon emissions in every part of our daily lives.

 Magnetic earth　　　　　　　　　　　　　　　　　**1600**

William Gilbert was a Cambridge scholar, natural philosopher and physicist from Colchester, credited by many as being 'the father of electricity and magnetism'. Britain was a dominant seafaring nation during Gilberts lifetime and mariners relied a great deal on the magnetic compass for Navigation. There was much mystery about where magnetism came from up to this point, with some eminent sailors such as Columbus thinking that the pole star affected the compass needle. This prompted Gilbert to carry out experiments in an attempt to clarify the phenomenon of magnetism once and for all. During his experiments, he noticed that magnetic forces frequently created circular motions, pointing to a link between this phenomenon and the Earth's rotation. This provided Gilbert with a theoretical basis for further research into the science of magnetism.

Physics

Magnifying glass   1250
Roger Bacon was a philosopher and inventor from Somerset. He graduated at Oxford University at the age of 13 and later became a master at his college. In the course of his many experiments, he was the first person to invent and document the magnifying glass, whilst researching the eye, reflected vision, refraction lenses and mirrors. Bacons invention happened long before the patent system was established and was therefore freely copied and enjoyed by many for years to come.

Mass spectrograph   1919
Scientist Sir William Aston is credited with building the world's first fully functional mass spectrograph in 1919. Aston was continuing his work with renowned scientist J.J. Thomson and during Aston's experiments with the fully built device, he was able identify isotopes in chlorine, krypton and bromine, providing positive evidence that these natural elements are made up of a mixture of isotopes. Aston's use of electromagnetic concentration in his Mass Spectrograph, allowed him to quickly detect at least 212 out of the 287 naturally occurring isotopes. The mass spectrograph is still in use today, allowing crime laboratories, water authorities and science researchers worldwide, when there is a need to identify a substance that may contain many other different elements.

Mercury vapor lamp   1901
A mercury vapor lamp is a gas discharge lamp that uses an excited state of mercury to produce a bright white light. Its inventor, Peter Cooper Hewitt, patented the mercury vapor arc lamp in 1901.

Microwave   1864
James Clerk Maxwell was a brilliant inventor who discovered the existence of microwaves, by the use of and proof through his famous paper of 1864, using proven equations and calculations. Based on his research, 24 years later, Heinrich Hertz proved Maxwell's claims by constructing a device that detected microwave radiation. Many scientists believe that Microwave discovery is possibly one of the greatest inventions of the 19th Century.

Micro switch or miniature snap-action switch 1932

The miniature snap-action switch or as it was trademarked 'micro-switch', is actuated with minimal force, through the use of a tilting device. Most common uses of the micro switch include elevator safety switches and microwave oven doors. Its inventor was Phillip Kenneth McGall of Freeport, Illinois.

 Motion - law 1687

Isaac Newton was one of Britain's greatest mathematicians and scientists. During his latter years at University, Newton was publishing some of his ideas in a journal about his thoughts on motion, calling them the three laws of motion. First law: often called the law of inertia; an object at rest will continue at rest unless acted upon by an unbalanced force. All objects resist change in their state of motion. Second law: when force acts on a mass, then acceleration is produced. Greater force is needed to accelerate the mass, the greater the mass is. Third law: for every action, there is an equal and opposite reaction. This means that an object will get pushed back equally as hard as it is being pushed in the opposite direction.

 Multiple coil magnet 1831

American scientist Joseph Henry developed a multiple coil electromagnet, with all coils connected in parallel. This method of winding increased the electromagnets current, thus generating a stronger magnetic field.

Physics

 Nanowire battery 2007

Consisting of a stainless-steel anode, the nanowire battery is covered in silicon nanowires. It uses silicon specifically due to its ability to accumulate ten times more lithium than graphite. This phenomenon allows greater energy density on the steel anode, reducing the overall mass of the battery. The nanowire battery also charges and discharges faster than a conventional lithium battery due to the higher surface area. In principal, a nanowire battery could run a smart phone a week on a single charge versus current technology of around a day. The nanowire battery was co-invented by Stanford University professor Dr Yi Cui and his team of colleagues in 2007.

 Nickel - zinc battery 1900

Invented by Thomas Alva Edison and his team, the nickel-zinc battery is rechargeable that can be used for portable tools and other electrical devices. Edison was granted a patent for this battery on October 8th, 1900.

 **Neutron** 1932

James Chadwick was a scientist who in 1932 followed a field of research that led to the discovery of the neutron. He quickly went on to calculate its mass. He predicted that neutrons could develop into a major breakthrough in the fight against cancer. Chadwick was also involved in projects developing the atomic bomb during his war years.

 Nuclear power - generation station 1956

The world's first commercial nuclear power station was at Calder Hill, Cumbria, also known as Seascale. When it first opened it was a reason for great pride and celebration, signifying the beginning of the fledgling atomic age. The site was closed in 2003 and commenced decommissioning, a process that will take up to 100 years to complete. It was also the world's first Magnox reactor, so called because the fuel vessels were produced from magnesium alloy.

Quantum cascade laser   **1994**

Co-invented at Bell Laboratories by Alfred Y Cho, Claire F Gmachi, Federico Capasso, Deborah Sivco, Albert Hutchinson, and Alessandro Tredicucci in 1994, the quantum cascade laser is a slice of semiconductor matter about the size of a pin-head. Within this matter lies electrons that are controlled within levels of aluminum and gallium compounds that are nanometers thick. Electrons jump from one layer to the other and when they jump, they discharge photons of intense light. Hence the name, quantum cascade laser. A patent was granted for this invention on October 10th, 1995.

 Radiocarbon dating   **1949**

Invented by Willard F Libby in 1949, radiocarbon dating is a method of determining the naturally present radioscope carbon 14 that establishes the age of carbonaceous materials within a range of up to approximately 60,000 years.

 Radio direction finder   **1901**

Invented by physicist John Stone, a radio detection finder is a device that pinpoints the direction of a radio source. It is a helpful navigation aid for ships and aircraft because lower bandwidth radio waves can travel long distance over the horizon and this is an example of the earliest form of radio navigation. Stone was granted a patent on December 16th, 1902.

 Radio - first transmission and receiving   **1856**

British engineer David Edward Hughes is documented as the first person in the world to transmit and receive radio waves. He designed and developed the synchronous type-printing telegraph. This was 40 years before Marconi (1896) was credited with sending a wireless telegraph and 10 years before Mahlon Loomis of Washington USA, sent signals by radio (1866).

Physics

 ### Radio - theory of electromagnetism   1873

James Clerk Maxwell was a mathematician and scientist, who through his research in the fields of electricity and electro-magnetism pioneered the concept of the 'Electro-Magnetic field'. It was during this research that James Clerk Maxwell proved the existence of radio waves and this discovery generated notional connection, that light waves were also connected to this electrical phenomenon. Thereby discovering the existence of radio waves, paving the way for even greater discoveries.

 ### Reed switch   1936

Operated by a positive magnetic field, a reed switch is constructed with two ferromagnetic custom shaped blades or reeds, placed in an airtight glass tube, with a predetermined space between them. Reed switches are used as sensors, switches and built in to relays found in many games and toys. The reed switch inventor was W B Elwood, an employee at Bell Telephone Laboratories.

 ### Relay - electrical   1835

In its basic form, a relay is simply an electrical switch that can open and close under the control of another circuit or low current switch. Prominent American scientist and inventor Joseph Henry invented the relay that was operated by an electro-magnet switching on or off that in turn opened or closed contacts controlling another circuit. Relays are still used in machinery globally, including the majority of cars.

 ### Richter magnitude scale   1935

Co-Invented and developed by Beno Gutenberg and Charles Richter in 1935, the Richter scale uses a number to quantify the volume of seismic energy discharged by an earthquake. Like decibels, it is based on a logarithmic scale, by calculating the logarithm of the total horizontal amplitude of the largest peak of energy from zero, using a Wood-Anderson torsion seismometer output, a number from one to 10 then signifies the strength of an earthquake.

Selective laser sintering   **1989**

An additive rapid manufacturing technique that employs a high-power laser to weld small particles of metal, plastic, ceramic or metal into a 3-dimensional mass, selective laser sintering is the first example of multimedia 3D printing. Its inventor was University of Texas lecturer, Dr Carl Deckard in 1989.

 Solar power - first solar cell patent   **1888**

Edward Weston was a chemist from Shropshire, who was acknowledged for his work in the fields of electrochemical cells, electro-plating and most significantly, the 'Weston Cell'. His solar cell patent was described as 'apparatus for utilizing solar radiant energy' and lodged for patent approval in New Jersey during 1888. This early solar cell was very inefficient, at less than one per cent and it took more than fifty years before Russell Ohl, who patented the Silicon semiconductor solar cell in 1946, developed a serious alternative. Even solar cells made today still only operate at less than thirty per cent efficiency.

 Space pen   **1965**

Commonly known as the zero-gravity pen, the space pen uses pressurized ink cartridges that is claimed to be able to write at any angle, under water on wet or greasy surfaces and in a zero-gravity state. The space pen was invented and patented by Paul C Fisher in 1965 following two years of NASA testing and was subsequently used during the Apollo 7 space mission in 1968.

 Spectrum - heterogeneity of light   **1655**

Prolific inventor, scientist and physicist, Sir Isaac Newton carried out a number of experiments between 1642 and 1655 indicating that he was the first person to understand the rainbow, by refracting white light through a prism, thus transforming it into its constituent colors of red, orange, yellow, green, blue and violet

Physics

Speed of light   1728
British clergyman and amateur astrologer James Bradley discovered the 'aberration of light'. In its own right this phenomenon is not that significant, but it encompassed and allowed Bradley to prove beyond reasonable doubt, that the planet Earth was travelling through space and proved that the universe did not revolve around Earth. Furthermore, through this discovery he was also able to calculate that light travelled at 670,616,629 mph, incredibly within only 1% of the accepted figure agreed by scientists today.

Stepping switch   1888
The stepping switch is an electromechanical device that permits an input to be connected to any one of a number of possible outputs, by direct control of electrical pulses. Invented and patented by Almon Brown Strowger, major users of the stepping switch were telephone exchanges, used in routing calls.

Strain gauge   1936
The strain gauge was invented by a professor from the California Institute of Technology, known as Edward E Simmons. Simmons invention was a measuring device that was used to gauge the tension or strain of an object and as the object became distorted, causing its electrical resistance to change too. This was the first example of precision measurement of material strain.

Strobe light   1931
Commonly known as a strobe, the strobe light is an apparatus that generates variable flashes of light, determined by the operator varying its speed. Most strobe lights were used as a warning light or to calibrate revolving motors or discs but are now used extensively as part of stage light shows in concerts and night clubs. Its inventor, Harry Eugene Edgerton, developed the strobe to aid his study of moving matters.

Telescope - reflecting   **1668**

Sir Isaac Newton, the famous British scientist and inventor was a fellow at Cambridge University's Trinity College. He developed the idea of a reflecting telescope and made it reality in 1668. This was a huge step in the technology of telescopes, because it made astronomical observation far more precise, by using two concave mirrors inside the main tube, rather than lenses to produce a magnifying effect.

 Tesla coil   **1891**

Nikola Tesla was a prolific Serb-American inventor, most famous for his invention of alternating current. His creation, the Tesla coil, is a variant of resonant transformer and was used in his experiments with electrotherapy, lighting, X rays and high frequency AC phenomena, to name a few. The Tesla coil was also used as an early wireless point to point transmitter of electrical power through inductance.

 Thermistor   **1930**

Invented by Samuel Ruben, the thermistor is a type of resistor with a variable thermal resistance built in, wholly dependent on temperature.

 **Transistor**   **1947**

Used to amplify or switch electronic signals, a transistor is a semi-conductor. The transistor can provide amplification of a signal because it is designed to allow the controlled output power to be far greater than the controlling input power. William Shockley noted in 1947, that when Walter Brattain and John Bardeen of AT&T Bell Labs performed an experiment using two gold point contacts applied to a crystal of germanium, a signal was produced and the output was greater than the input. Shockley who was the manager of Bell Labs semi-conductor research group at the time, saw its potential and worked over the following months to construct the first point-contact transistor. Through this development, Shockley is considered to be the 'father of the transistor' and this ground-breaking device is now the primary basis of all contemporary electronic designs. In recognition of their invention of the transistor, Bardeen, Brattain and Shockley were awarded the 1956 Nobel prize in physics.

Physics

 ## Ultra violet – black light   1935
Invented by William H Byler, the Ultra-violet or black light is a device that emits electromagnetic radiation that falls within the soft UV radiation spectrum, emitting very little visible light. In a dark room it will highlight most white objects such as clothing, eyes or teeth.

 ## Voltmeter - digital   1953
An instrument used to measure electrical potential or voltage between two points, the Voltmeter comes in two variants: analog or digital. The analog meter is not so accurate as it uses a dial gauge with a needle. But the modern digital meter can measure, often up to 10 decimal places. The digital Volt meter was invented by Kaypro founder, Andrew Kay.

 ## Wind-up radio   1991
Trevor Baylis was watching a TV programme filmed in Africa about AIDS. During the documentary, the presenter commented that a sure-fire way to stop the spread of the disease would be for all people to have access to good information on the radio. It was at this point that Baylis designed a radio that required no batteries, instead, taking its power from a wind-up internal spring powered by a hand crank. He demonstrated a model to Nelson Mandela and ever since that meeting; his radios have been distributed all over Africa.

 ## Zener diode   1915
A type of diode that allows current through in the forward direction, similar to a normal diode, but the Zener diode also permits current to flow in the other direction if the voltage is higher than the breakdown voltage, known as the Zener knee voltage. The device was invented by Clarence Zener whilst investigating its strange electrical properties.

Zero - devised scale to show absolute zero   1848

It was not until 1848 that William Thompson or Lord Kelvin developed Robert Boyles original theory on absolute zero. Lord Kelvin decided that it would be helpful to be able to characterize extremely low temperatures precisely. The breakthrough came when he noted that molecules stopped moving at absolute zero. His rules were known as the Kelvin scale, where absolute zero is minus 273.15 degrees on the Celsius scale. His work is vitally important in the field of superconductivity, where superconductors are extremely efficient at conducting electricity only at low temperatures; discovered only after Kelvin's death.

"The art of medicine consists in amusing the patient while nature cures the disease"

Voltaire

Sport

Introduction to modern sports

The majority of modern sports can be traced back in their creation to Great Britain, not just in standardization and rules, but also their origins and development. Most significant of which are tennis, cricket, rugby, football, squash, hockey, golf, snooker, billiards, darts and table tennis. More sports can also trace their roots back to British sports, such as: American football (rugby) and baseball (rounder's). It's clear that not every ball game was originally created in Great Britain, but what did happen was the establishment of British standardization in both rules and structure for most of the world's most competitive and popular sports in the world today.

American football 1869

Known affectionately outside the USA as American football, football is a spectator sport based on competitive physical play and strategy. Each team can score points by advancing the ball to the opposite team's end zone. Contrary to the sports name. unlike the British game of football, the US game relies mainly on carrying and throwing the ball, rather than foot use. Although the games origins date back to the British game of rugby; the sport, its codified rules and pitch layout is generally credited to its creator, Walter Camp and these base rules eventually developed and led to the hugely popular game that it is in the USA today. The first professional game of football was played between Princeton University and Rutgers University on November 6th, 1869.

 ## Artificial turf 1960

A man-made synthetic grass or turf designed to look like real grass and most often laid and used in both indoor and outdoor sports stadiums and arenas. Artificial turf was invented and developed by David Chaney and his creation was first installed at the Reliant Astrodome in Houston, Texas.

 ## Badminton - racquet sport 1873

Badminton was first played in the form that we know today, at a party held by the Duke of Beaufort at Badminton House in Gloucestershire. Its origin in its most basic form probably dates back more than 2000 years. However, in the 1600s battledore and shuttlecock was an upper-class pastime in England that spread to many European countries. Badminton was played by two people hitting a shuttlecock back and forth, without allowing it to hit the ground.

In India, a net was introduced to the game by British officers in the 1800s, to add more of a challenge to their pastime game; then called poon. When officers brought this game back to England, it was introduced to the Duke of Beaufort during 1873 at his home in the stately manor of 'Badminton' in Gloucestershire, where the game went on to spread and become more popular.

Sport

In 1898, the first open tournament, now known as Badminton, was held at Guilford in the world's first 'all England' championships. During the 1930s the USA, Denmark and Canada all became passionate followers of the game. In 1934, the International Badminton Federation (IBF) was formed. Founding members were, England, Wales, Scotland, Ireland, France, New Zealand, Holland, Canada and Denmark. India joined as an affiliate in 1936.

 **Baseball** **1755**

Documented by William Bray's diary entry. It is now confirmed that local historians from Surrey have found conclusive evidence that baseball was played in Britain more than 20 years before American independence. The diary belonging to local lawyer and diarist William Bray has now been attributed as the first description and mention of the game of baseball. It describes the game being played in Guilford by men and women in March 1755 on Easter Monday.

 Baseball – codified rules **1845**

The USA's national pastime and sport, baseball was formalized and codified by Alexander Cartwright, originally known as the knickerbocker rules. At the time, New Yorker Cartwright's 14 entry rules were similar to the British game of rounders on which he based his game, with exceptions such as the base layout in the US game, which is set in a diamond formation rather than the square field used in the British game of rounders.

 **Basketball** **1891**

The game of basketball is a sport invented by British doctor, James Naismith and first played in 1891. In the UK it later evolved into the game of netball, now predominantly played by women, except in the USA where basketball is now principally played by men's teams.

Billiards   19th Century

The game of billiards is generally played by two players or teams and originates in Great Britain, spreading rapidly to the colonies during the 19th Century. There are two cue balls in billiards, both white, but one is marked with a black dot and a red object ball is used. The game is played on a table of similar size and dimensions as a snooker table. There is a different cue ball assigned to each player or team and points are scored for cannoning and pocketing the balls in any one of the six pockets around the edge of the table.

Bowls - the sport   1299

The British sport of bowls has been traced back to the 13th Century. The world's oldest bowling green still played on, is in Southampton. Records reveal that it has been functioning in its current form since 1299. The game is now played globally.

 Boxing - Queensbury rules   1865

John Chambers was the Marquis of Queensbury, who publicly sanctioned the world standard code of boxing, written by John Graham Chambers. The code is still the basis for professional boxing rules used to this day. The rules were the first to mention the compulsory wearing of gloves and designed to be used for both amateur and professional boxing.

Chair lift - ski   1936

An engineer from the Union Pacific Railroad, John Curran, invented and built the first chair lift. His design consisted of a continuously circulating steel cable, positioned between two locating towers. The ropes hung a series of chairs that were equally spaced and these typically went up a hill to the top terminal or tower and then back down to the lower terminal. Curran's first working example was installed at the Proctor Mountain resort in Sun Valley, Idaho. The same principle is now used in most ski resorts around the world.

Sport

 Cricket - first rules   **1774**

Cricket, the world's second most popular sport (the British game of football or soccer, is the world's most popular sport). Cricket originated in rural England in the 16th Century and is now played in most countries across the world. Often derided and frowned upon, the world's first international cricket match was played between the USA and Canada in 1844. Now, according to the ICC (International Cricket Council) registrations, more than 107 countries are registered and actively compete and play professional cricket.

 Darts - traditional pub game   **1896**

Brian Gamlin devised the dartboard numbering layout. Some form of the game of darts has been in existence for many hundreds of years in Great Britain, but it was Brian Gamlin a British carpenter from Lancashire, who developed the numbering layout that is still used to this day. He devised the seemingly random layout of numbered segments to punish inaccuracy from players.

   **1863**

The Football Association (FA) was formalized in 1863, although it flourished in England from the 8th century, but under the more recent association directive from the 'Cambridge rules'. The FA was formed during an historic meeting at the freemason's tavern in London's Great Queen Street, chaired by London solicitor Ebenezer Morley, who was a huge advocate of formalizing the rules of football, similar to the way Cricket rules had been formalized. At the meeting with Morley, were 12 London area football club representatives, there for the sole purpose of determining a strict code of rules to help regulate the game.

The only surviving club from that meeting is Civil Service FC, a member of one of the lower southern league amateur divisions. The FA is the world's oldest and original football association. Football is the world's most popular sport by a huge margin.

Formula 1 - motor racing   1946

Formula 1 is a global racing car sport to which all participants must comply with a strict set of rules, hence the prefix formula. The sport known as F1 was planned in the 1930's, buoyed on by European success in motor sport, but was put on hold due to the onset of WWII. The concept was resurrected in 1946 and the first world championship race took place in May 1950 at Britain's Silverstone racetrack. F1 is now the most popular and richest motor sport in the world and is headquartered in Great Britain. Three quarters of F1 teams design and build their cars in Britain.

 ## Figure skating   1772

The world's first ice skating association was formed in Britain in 1772. The art of figure skating involves either singles or pairs of ice skaters performing jumps or spins, footwork and lifts, in a style akin to ballet dancing. Briton Robert Jones wrote a paper on the art in 1772 called 'A treatise on skating'. The sport has now developed worldwide and has been an Olympic event since 1908.

 ## Golf   15th century

There is and always will be much debate about the origins of the game of golf. However, the modern-day game can be traced back to its British origins in the 15th Century. It was known as the second game of kings with King James IV documented as playing the game regularly in 1500 and Mary Queen of Scots played a game just prior to the death of her husband Lord Darnley. King James VI allegedly played a game shortly after ascending to the throne at Blackheath in London in 1603. The Honourable Company of Edinburgh Golfers established the first standardized rules of modern golf and the oldest golf club in the world is St Andrews (1754). Golf is now the 10th most popular international sport in the world.

Hockey   19th century

The game known as hockey, was played in English private schools in the early 1800s. In 1886 the Hockey association was formed with seven London clubs joining and at that point, rules were drawn up to form the basis for the modern game that we know today. Hockey spread throughout the British Empire with the advent of its popularity amongst the British Army. The International Hockey Federation (IHF) formed in 1924 and field hockey was introduced for the first time as a men's only game at the 1924 summer Olympics.

Horse racing - thoroughbred   1750

In Britain, horse racing is the second biggest spectator sport in the country, next to football. It is one of Britain's oldest sports and has been established since Roman times. Most of the sports rules, regulations and traditions have been developed in Britain, including the establishment of the Jockey Club in 1750, who created the rules of racing. Britain has been the world centre for thoroughbred racing for many centuries and boasts some of the world's most famous racecourses such as Ascot, Cheltenham, Windsor and Epsom. There are two forms of horseracing in Britain: flat racing, run over distances of up to two miles, 5 furlongs and 159 yards and National Hunt Racing: run over distances of up to four and a half miles with horses jumping over hurdles or fences. The Cheltenham Festival is the world's number one jump racing festival.

Ice hockey   1818

Contrary to common belief, Charles Darwin observed ice hockey and noted in a letter to his son William whilst reminiscing, that he played hockey on an 'iced over pond' on 'ice skates' at school. He attended Shrewsbury School between 1818 and 1825. This was revealed in a book called 'Origin of Hockey" by two Swedish colleagues Carl Gilden, Patrick Houda and French-Canadian Jean-Patrice Martel in 2014.

Netball   1890s

The British sport of netball emerged from early versions of another British game called basketball; a sport invented by British doctor James Naismith and first played in 1891. The game of netball developed shortly after basketball and was initially played by women. Although not technically intended, the game of netball is still dominated by female players the world over.

Olympic games - London   1908-2012

London is the only city to have hosted the summer Olympic games three times: in 1908, 1948 and 2012. Despite Britain covering less than 1% of the world's landmass, it ranked third in the world in the 2012 medals table, just behind the USA and China. But the best was yet to come; at the 2016 Olympics in Rio, Britain ranked an outstanding second overall in the medals table behind the USA.

Paralympic games   1948

The first Paralympics games for disabled athletes were held in London in 1948. German born Dr Ludwig Guttmann from Stoke Mandeville hospital, held a competition for war veterans with spinal cord injuries. The games were originally called the 'Wheelchair Games' and were purposely planned to run the same summer as the 1948 Olympic games. The games were held again in 1952, this time with Dutch veterans taking part alongside the British participants. It became the first international competition of its kind. Then in 1960 and no longer the preserve of war veterans, the world's first official Paralympics games were held and attended by 400 athletes from 23 different countries.

Polo   19th Century

The game of polo is a team sport played on horseback, formalized and popularized in Britain in the early nineteenth century. The aim of the game is to score points or goals against the opposing team by hitting a small wooden or plastic ball into the other team's goal, hit with a long-handled mallet. The polo pitch is traditionally 300 yards long by 160 yards wide and each team has four riders and horses. The professional sport of polo is played in 16 countries.

 Racquetball   1968

A racquet sport played with a hollow rubber ball, racquetball, can be played on both on an outdoor and indoor court. A professional tennis player, Joseph Sobek is credited with inventing the sport of racquetball. He wanted to create a fast-paced sport that was not too difficult to learn or play and so he set to work devising a codified set of rules and constructed his first strung paddle, calling his game 'paddle rackets'.

 Rounders   15th Century

The game of rounder's is the origin for the game now known as baseball, but rounders has been played in Britain since Tudor times. It is a bat and ball game using a hard leather ball, similar in size to a cricket ball hit by a mildly tapered tubular wooden bat. The team members can score by running around the four bases on the pitch. One full round counts as one point. The game is very popular among British and Irish school children. In the USA, adults play the very similar game of baseball professionally.

Rugby league   1895

The game of rugby league was formed as a separate game from rugby union following a disagreement of paying players. Its history dates to 1895, when a split in rugby union teams resulted in the formation of rugby league paid professional players, and rugby union non-paid amateurs. Slowly, the two games evolved into distinctively separate games. Rugby league in Britain is mainly a northern game consisting of 13 players.

Rugby union   1823

Legend has it, that British Public Schoolboy William Webb Ellis started this international sport while playing football at Rugby School during a match one day. In the heat of the moment, he picked the ball up and ran with it into the goal line. The rest is history. Rugby union is now played with 15 players and is the main sport in both New Zealand and South Africa and played enthusiastically worldwide.

Rugby union was completely amateur and non-salaried until a split and argument over payment of players in 1895 erupted. This dispute led to the split of rugby union and formation of rugby league, where all players were paid professionally and from that point on, the two games have evolved with two different sets of rules. The embargo of paying rugby union players was not lifted until 1995 and only then was professionalism sanctioned by the games governing body. A derivative of this game is American football, although the players of rugby union and league players do not wear the body protection that US American footballers don.

Sailboard and sail boarding   1958
Briton, Peter Chilvers built the world's first sailboard. Chilvers was a 12-year-old boy from Hayling Island in Essex at the time of his invention. As a child, he enjoyed most water sports and one day while at the beach near his home, his inquisitiveness got the better of him and he mounted a sail on his surfboard. Chilvers became heavily involved in developing and getting the sport internationally recognised. The sport grew in popularity so fast around the world that it became an Olympic sport in 1984.

Snooker - origin   Mid 19th Century
Colonel Sir Neville Chamberlain bestowed the name 'Snooker' on the game while serving in the British Army in India. He commented that a young player was 'a real snooker', a common term given to a first-year cadet, referring to their inexperience at the time. Snooker is a derivation of the original table game of billiards, a sport with just two cue balls and an object ball. However, using the same size table, snooker uses 15 red balls, six colored balls and a white cue ball.

Snowboarding   1965
A sport that comprises decent down a snow-covered slope on a snowboard fastened to the rider's feet, snowboarding has its origins in skateboarding and surfing. Sherman Poppen invented and rode the first snowboard in 1965 and it is now a winter Olympic sport.

Softball   1887

In the sports category of bat and ball games, softball is a variant of British rounders and baseball. The main differences are that softball requires a smaller playing area and uses larger softer balls. The game started life indoors in Chicago and was originally invented by George Hancock in 1887.

Squash   1820

The game of squash was first devised and played at British public-school Harrow, in around 1820. Pupils discovered that a deflated tennis ball squashed on contact with a wall and this was the catalyst for a game that developed a larger selection of shots, that needed more effort for the players, rather than waiting for a ball to bounce in simple tennis. Squash is now an international and Olympic sport played by millions of players worldwide.

 Stock car racing   1936

A form of automobile racing, stock car racing uses short oval shaped tracks. Some are just dirt tracks and longer oval tracks are commonly known as speedways. The inventor credited with this sport was Sig Haugdahl and his very first race on March 8th, 1936 was just for family sedans.

Surfboard fin   1935

Invented by surfer Tom Blake, the surfboard fin is attached to the rear bottom end of a surfboard to help stabilize, steer and prevent the board spinning over. Blakes invention was effectively a modern development on a centuries old design, that resulted in spawning a whole new international sport to this day.

Swim fins - flippers   **1717**

Unlike todays versions of plasticized swim fins, prolific innovator and inventor Benjamin Franklin invented a large wooden type, shaped like a painter's palette, designed originally to use with a swimmer's hands. His design provided the swimmers arm stroke with greater speed. In 1968, Benjamin Franklin was posthumously inducted into the International swimming hall of fame.

 Table tennis or ping-pong   **1880s**

The history of ping-pong or table tennis originated in Britain and probably evolved as an indoor version of the British game 'royal tennis' or real tennis, widely played in the Medieval era. There are also suggestions that the game was once played by British Army officers in India and South Africa and known as indoor tennis, using the lids from cigar boxes and balls molded from cork wine stoppers. But in 1900, Briton James Gibb brought back hollow celluloid balls to Britain from the USA, in the form that are now used for the sport of table tennis the world over. In 1901 both the Table Tennis Association and the Ping-Pong Association were formed in Britain.

 Tennis   **1874**

British Solicitor, Major Harry Gem from Birmingham, devised the sport of tennis with a Spanish friend and two doctors in 1874. Gem then founded the world's first lawn tennis club: the Leamington Tennis Club. Tennis is now the world's fifth most popular sport in the world. Two or four players can play the game either singly or in pairs, played predominantly on a grass court, or in some warmer climates, on clay-based courts.

 Trampoline   **1934**

A gymnastic and fitness apparatus, the trampoline comprises a sheet of strong fabric, strained across a steel framework attached with numerous coiled springs between sheet and frame. This tensioned sheet then thrusts the jumper high above the sheet. A version of what is now considered the modern trampoline was invented by Larry and George Nissen in 1934.

Sport

 ## Volleyball   1895

An Olympic sport, in which two teams of six participants play against each other, separated by a high net. The ball is knocked over from one side to the other until the ball either goes out of play or it lands on the court, in which case a point is conceded. Later named Volleyball by Alfred S Halstead, it was originally known as 'mintonnette' and was invented by William G Morgan while he was studying at a YMCA in Holyoke , Massachusetts.

 ## Water skiing   1922

Invented and developed by Ralph Samuelson when he used two small surfboard like wooden planks strapped to his feet and a clothes line as a tow rope, tied to the rear of a motor boat on Lake Pepin in Lake City, Minnesota. Samuelson became very competent and took to the road, touring the country demonstrating his water-skiing stunts. It wasn't until 1966 that the American Water Ski Association finally acknowledged Samuelson as the pioneer of the sport of water skiing.

Transport

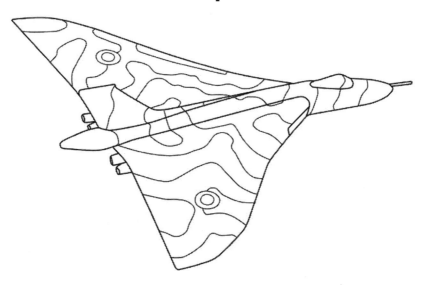

Aviation

✸ Aerial refueling   1935

British aviation scientist Sir Alan Cobham was a born in 1894 in London. His illustrious career spanned WWI as a pilot and he once flew to Melbourne Australia from Britain in a De Havilland DH50 Floatplane. In his pioneering work on air-to-air refueling, Cobham's early research was based on specially adapted Airspeed Courier planes. He successfully developed and mastered the in-air refueling technique during the latter part of WWII and his design was successfully taken up by the RAF and US air forces. His old company Cobham PLC is still active. This invention made possible the world's longest non-stop mission for a bomber by the RAF's Avro Vulcan, during the Falklands conflict, flying 8000 miles.

Aviation

Ailerons 1868

The aileron was first developed and patented by scientist, Matthew Watt Boulton in 1868. The patent was based on his academic paper 'On aerial locomotion'. Then 38 years later, the US patent office granted a comprehensive patent to the Wright brothers for their identical version of Bourton's design called 'A system of aerodynamic control'. Substantial litigation ensued over the legal aspects of lateral roll control. The US government eventually passed a legal resolution in the Wright brothers favour in the US, just before the onset of WWI.

Aircraft - first powered flight 1848

John Stringfellow was a British inventor who travelled by air at just under 10 metres. His contraption was then known as the 'Ariel steam carriage'. He was based in Sheffield before moving to Chard in Somerset to pursue a career in the lace industry but had always dreamt of starting an intercontinental airline. Stringfellow set up a business with an engineer called William Henson to fulfil this dream.

Unfortunately, despite their best efforts at producing glossy publicity brochures and holding many pre-sales forums, they failed to attract any financial backers who were interested in working with them.

Henson parted company at this stage, but Stringfellow persevered and designed optimal wing designs and new lightweight construction methods, whilst simultaneously calculating wing lift to weight ratios for his various wing shapes. Eventually he built a working test model using optimal wing design and powered by a twin propeller steam engine. The frame of the aircraft was extremely delicate due to the need to make a lightweight structure to ensure lift. The vulnerability of the silk covered wings and bamboo frame struts meant that it could not be tested outdoors as the silk became quickly sodden with moisture in an external environment.

Consequently, initial tests took place inside a large Chard based milk mill in Somerset. Eventually in 1848, his unmanned, powered flying machine lifted off the mill floor at speed and travelled a mighty 10 yards (just over

9 metres) and it became the world's first powered flight in a fixed wing aircraft. A model of this historic aircraft is now on display at the Science Museum in London.

Aircraft - first manned heavier than air flight   1849
Sir George Cayley (the 'father of aerodynamics') was a British inventor who designed and built a triple winged glider in Brompton Yorkshire, carrying a 10-year-old boy. This feat became the world's first manned flight by a heavier than air craft. Sir George was a prominent MP and extraordinary inventor who also worked on the development of caterpillar tracks, early seat belts and lifeboats.

Aircraft - first manned controlled flight   1853
Sir George Cayley was the first inventor to shift from the idea that all aircraft needed wings like a bird. He designed and built the world's first full sized controllable glider, capable of carrying a fully-grown adult and it was test flown by his coachman over Brompton Dale in Yorkshire in 1853. On landing, records suggest that his coachman immediately tendered his resignation.

Aircraft - first manned powered controlled flight   1903
It is acknowledged that the Wright brothers were clearly not the originators of fixed wing heavier than air flight, nor sole masters in aeronautic theory and design. Though, what is absolutely certain is that they can lay claim with confidence, to being first to demonstrate a heavier than air flying machine that was powered in a sustained flight under the control of a pilot. The Wright brothers secret breakthrough aside from dogged determination and perseverance, was their development of 'three axis' flight control system that enabled the pilot to effectively steer the aircraft whilst maintaining its balance. Variations of this Wright brothers flight control principal then became standard on fixed wing aircraft in the ensuing decade. The brothers were subsequently granted a patent for 'Kitty Hawk' the aircraft, in May 1906. The rest is history.

Aviation

 Aircraft - first powered controlled flight   1899

Percy Pilcher was a British inventor who set an unpowered gliding record of 250 metres in Eynsford, Kent and came extremely close to becoming world famous by upstaging the Wright brothers record by 13 years! Following his record manned glide, he then focused on powering his 'hawk' glider and decided on developing a triplane three-winged design, because the additional weight of the third wing did not add significantly to the overall weight of the craft but did make a huge difference to its lift. He used a small petrol-powered internal combustion engine to power the aircraft.

On September 30th, 1899 he arranged a demonstration of his new powered aircraft at a venue near the town of Rugby. Unfortunately, the craft was not flight ready. The day was stormy and very windy and rather than disappoint the invited crowd, he decided to fly his glider, the Hawk. In the event, the aircraft proved very difficult to control in the gusty conditions and the tail plane snapped off while flying, sending the craft plunging to the ground, killing Percy Pilcher in the process.

Unfortunately, after his untimely death, although almost complete and ready to fly, work was halted on his powered aircraft and dismantled for scrap. However, in recognition of Pilcher's achievements, a BBC programme team worked with Cranfield University in 2002, to build a replica of his powered tri-plane to prove it could fly. Once built, it flew for more than a minute and proved a far better aircraft than the Wright brothers 'Flyer'.

 Aircraft - landing on a moving ship   1917

Squadron Commander E.H. Dunning of the Royal Naval Air Service made the world's first aircraft landing on a moving ship. He landed his Sopwith Pup on to the deck of HMS Furious in Scapa Flow, Orkney on 2nd August of 1917. Alas, five days later he was killed while attempting a second landing on the ship that day; his wing tip was caught by an updraft, tossing his plane off the port side of the ship, rendering Dunning unconscious and he drowned, unable to break out of the cockpit.

Aircraft - first turboprop airliner　　　　　　　　1948

British designed and built; in 1948 the Vickers Viscount was the world's first turboprop airliner to enter commercial service. Within a few years, more than 450 were built and sold worldwide to many countries including the USA and the Far East.

 ### Aircraft - first VTOL　　　　　　　　　　　　　　　1954

The Rolls Royce thrust measuring rig (TMR), made the world's first vertical take-off and landing (VTOL) in this year. It was nicknamed the flying bedstead. This development led to the first vertical take-off and landing engines used in the British made Short SC.1 in 1957, using four engines for lift and one for forward thrust. However, this aircraft, although a capable VTOL craft, had very limited flight range due to its five engines.

 ### Aircraft - vertical take-off and landing　　　　　　1965

The British designed and built Hawker Siddeley Harrier jump jet is the World's first commercial and most successful VTOL (vertical take-off and landing) aircraft ever made. The Harrier was the first VTOL to use a single engine (Bristol Siddeley Pegasus) with pivoting nozzles that directed thrust for both forward and vertical flight.

The Harrier is still in service worldwide, primarily flown in short take off and vertical landing mode (STOVL) that enables the craft to carry a higher weapon or fuel load. Designed and launched commercially, initially for the British Navy whose fleet was fully retired in 2010. The United States Marine Corps and the Spanish and Italian navy's still use the AV-8B Harrier 11.

Aviation

 ### Aircraft - first commercial supersonic airliner   1969

The Anglo-French Concorde broke the sound barrier and accelerated to Mach 2 (twice the speed of sound). Concorde employed the world's most powerful jet engines in a commercial passenger aircraft, taking it to speeds of up to 1,350 mph and it entered full commercial service in Great Britain on January 1976 between London and Bahrain.

The Rolls Royce Olympus engines each used 'reheat' technology, adding fuel to the final stage of the engine, producing the additional power needed for take-off and the acceleration up to and beyond supersonic flight.

Concorde's fastest transatlantic flight was on 7th February 1996, completing the New York to London flight in 2 hours 52 minutes and 59 seconds. Concorde only ever saw service with two airlines – British Airways and Air France. BA withdrew Concorde in 2003 following the fatal Air France Concorde crash in Paris and the remaining BA fleet was retired for posterity and preservation at airfields in Barbados, Edinburgh, Filton (Bristol), Manchester, New York, Seattle and Heathrow.

 ### Aircraft - first autonomous aircraft   2009

The Mantis, thinks for itself, decides for itself and flies by itself. The mantis is a twin engined unmanned aircraft designed and built by BAE systems for the UK Ministry of Defence, developed from concept to first flight in just 19 months. Its first flight was on October 21st, 2009 at the Woomera test range in South Australia. It has a wingspan of 22 metres and uses 2 Rolls Royce M250B-17 turbo shaft engines.

Mantis has a maximum speed of 345mph and can cruise at 230 mph. Its maximum endurance is 30 hours per flight. The Mantis is designed to be used for both reconnaissance and attack operations and can be armed with six Hardpoint missiles. It is the world's first plane to be able to pilot itself and plot its own course, whilst communicating with ground staff to share its observations and strategic data.

Air force - world's oldest Air force   1918

Britain's RAF is the world's oldest independent air force, being formed on April 1st, 1918. At the time of its formation, it was the largest air force in the world and it is most famous for its role in its successful campaign of the battle of Britain, fighting the huge forces of Nazi Germany under the command of Hitler. The German Luftwaffe hugely outnumbered the RAF, yet the battle was won on courage, skill and superior aircraft.

 ### Autopilot   1912

An electrical, hydraulic or mechanical process that is used to specifically direct aircraft but can also be used on boats, rockets or road vehicles, is called an autopilot. Inventor Lawrence Sperry first demonstrated an aircraft autopilot system in 1914 and it proved so successful that the pilot was able to fly the aircraft without his hands operating the controls.

 ### Ejector seat   1940

Sir James Martin was an aeronautical engineer who founded 'Martins Aircraft Works' in 1929. He forged a close friendship with test pilot, Captain Valentine Baker. On 12th December, Captain Baker was tragically killed during a test flight of the Martin-Baker MB3 prototype. His death affected Martin so much, that his focus immediately turned to pilot safety. In 1944, Britain's Ministry of Aircraft Production invited Martin to investigate the practicality of developing a means of assisted escape for pilots in all their fighter planes. After many designs, he decided that the most practicable and viable scheme was that of a forced ejection of the seat with the pilot still strapped in place. He decided that the best method for this was to use an explosive charge. The modern-day ejector seat was born and the Martin Baker ejector seat design is now used the world over and to date has saved the lives of more than 7000 pilots and aircrew.

Aviation

 Head-up display 1942

Britain's Royal Aircraft Establishment designed the first prototype head-up display that was built by Cintel. The first system was successfully installed into the famous fighter/bomber, the Blackburn Buccaneer in 1958. The same basic design is still used in modern fighter aircraft and is now even offered as an option in some prestige cars.

 International daily air passenger service 1919

The world's first daily air international passenger service started life on a grass airstrip in Hounslow Heath, just a few miles from present day Heathrow Airport. The service ran daily between London and Paris using converted De Haviland bombers from WW1. The air service was run by Air Transport and Travel Ltd and the airline was named, Imperial Airways, ultimately becoming British Airways in later years. The Hounslow Heath Airport boasted the world's first arrivals and departure halls and the largest control tower in the world at the time.

 Jet airliner - world's first 1949

The De Havilland Comet was a revolutionary new streamlined jet airliner, the like of which the world had never seen before. The Comet had its inaugural flight in 1949, featuring a pressurized fuselage for safe high-altitude cruising. Each Comet came with 4 De-Havilland Ghost turbojet engines hidden in the wings. Several versions of this aircraft were made and the most radical version was the Hawker Siddeley Nimrod, the world's first Jet maritime reconnaissance aircraft. Nimrod saw service in the RAF until 2011, more than 60 years after the Comets first flight.

 Radio altimeter 1924

Invented by engineer Lloyd Espenschied, the radio altimeter measures altitude above the ground that is currently beneath a rocket or aircraft. The radio altimeter measures the distance via radio signals and the time they take to bounce back, regardless of the height of ground above sea level, contrary to a barometer that relies only on barometric pressure.

GB & USA: The Mothers of Invention

 SABRE HOTOL - hypersonic jet/rocket circa 2020

The SABRE (Synergistic Air-Breathing Rocket Engine) is the world's first hypersonic jet/rocket capable of working safely in space, allowing the opportunity of HOTOL (Horizontal Take Off and Landing). SABRE is still at the conceptual stage, but under development and early testing by British company Reaction Engines Ltd. SABRE is designed to propel an HOTOL craft at speeds in excess of Mach 5 at an altitude in excess of 28.5km high above the earth.

 Stealth aircraft 1981

The world's first operational stealth aircraft was the Lockheed F-117 Nighthawk. Designed around stealth technology that made it almost invisible to conventional radar, its maiden flight took place in 1981 and saw active operational service in 1983.

 Supersonic aircraft 1947

The sound barrier, in aviation terms, generally refers to the point at which an aircraft thrusts from subsonic to supersonic speed. US air force captain, Chuck Yeager claims the breakthrough into supersonic air travel, when he broke the sound barrier in 1947 flying supersonic in a Bell X-1.

 Tiltrotor 1930

Invented and developed by George Lehberger, the tiltrotor was an aircraft that used two powered rotors attached to revolving shafts at the ends of two fixed wings of an aircraft. These rotors provided lift and propulsion, by combining the vertical lift of a helicopter with the range and speed of a conventional propeller powered fixed wing aircraft. The rotors were fixed to a ninety-degree tilt device enabling the engine and rotor to adopt a conventional propeller stance for normal flying at height, then revert back to rotor position for a vertical landing. Lehberger was granted a patent in September 1930.

Aviation

 ## Wide body aircraft   1969

A large airliner with two passenger isles running down the length of the aircraft is known as a wide body aircraft. The world's first wide body aircraft to enter service was the Boeing 747, affectionately known as the Jumbo Jet. For 35 years, the 747 ruled the skies as the world's largest passenger airliner until the larger Airbus A380 had its maiden flight in 2005. The Boeing company and its design team are credited with inventing the world's first wide bodied aircraft.

 ## Wind tunnel   1871

British aeronautical engineer, Francis Herbert Wenham realized that in order to test all the forces and performance of good wing and aircraft design, there needed to be a method to test them first. In Great Britain, the science of aeronautics was developing fast and 50 years before the Wright brothers well-documented flight, British inventors had already designed and tested powered and controlled flight.

Although the value of a wind tunnel is obvious to us today, it was not created before test flights started in earnest. Wenham realized that to advance the science of flight and most importantly its safety, an alternative was needed to the 'finger in the air' testing of old. He had tried a similar design of a whirling arm as sir George Cayley used, but after several disappointments, he convinced the Aeronautical Society to build the world's first wind tunnel. The design was a long 12-foot tube, 18 inches square, with a steam driven fan, blowing air at one end down the tube, to the model aircraft at the other end. The testing went better than expected and Wenham's wind tunnel design was of huge importance to the pioneers of the fledgling British aircraft industry.

 Winglets 1979

Now a fixture on almost all commercial airliners, winglets are typically designed to increase the effectiveness and efficiency of fixed wing aircraft. Aeronautical engineer, Richard Whitcomb set out to improve a concept that dates back to the 19th century suggesting that winglets could improve stability and economy, using the same principal as the wing tips of birds, particularly birds of prey. His development work was a huge success and the first aircraft to be fitted with winglets was a military refueling tanker, the Boeing KC-135 Stratotanker on July 24th, 1979.

Wing warping 1911

The concept of wing warping involves the twisting of an aircraft's wings to produce sideways or lateral control and involves the entire wings structure to twist slightly in a circular motion. The invention of wing warping is attributed to Wilbur Wright who conceived the idea and concluded that the roll of an aircraft could be controlled by the movement of the aircraft's wings. This was the predecessor to the aileron that was invented by French aviator, Henry Farman.

> "I am well convinced that Aerial Navigation will form a most prominent feature in the progress of civilisation"

George Cayley 1849

Rail

 Flanged T-rail   **1831**

The flanged T rail was invented and developed by Robert L Stevens who worked for the Camden & Amboy Railroad Transportation Company. The flanged T rail was constructed from iron with a flat base in an inverted T configuration, thus there was no requirement to clamp the rails upright. Examples of Stevens invention can still be found in America today.

 Knuckle coupler   **1873**

Designed as a by-product of a standard coach coupling device on railroads, the knuckle coupler had a draw head and revolving hook, much the same as a human knuckle joint operates, hence the name. Eli H Janney's invention and subsequent patent became the standard coupler for the remainder of the 19th century.

MAGLEV - world's first MAGLEV Train   1984

Birmingham International Airport was the world's first site to use a passenger train in commercial service, solely using the MAGLEV system of propulsion. Maglev is derived from MAGnetic LEVitation, a means of propulsion that thrusts a vehicle by the use of magnets instead of conventional wheels. By using an electromagnetic central mono track, an object or vehicle can be lifted and propelled dependent on the metal type offered to the magnetized rail. Attached underneath the vehicle, some metals lift and propel the mass forwards, while others act in the opposite direction and are employed as brakes or to reverse the vehicle, dependent on the polar direction of the rail.

The MAGLEV principle was invented and developed by British scientist, professor Eric Laithwaite. The main advantage of MAGLEV systems is that because the drag and friction of each vehicle is low, the vehicles can travel much faster in a safe and controlled manner. Whereas the negative aspects are that maintenance is very high and because the system is not efficient in its energy consumption, most MAGLEV systems have now been decommissioned.

Today, there are only two commercial MAGLEV schemes that are still in operation: Shanghai, China and Aichi, Japan. Both are very controversial due to the huge cost of construction, maintaining and running the systems.

 Pantograph   1904

Invented by John Q Brown of the Key Systems shops for their commuter trains that ran between East bay to San Francisco in California, the diamond shaped pantograph was inspirational. The Pantograph was sprung loaded so that it was always expanded as far as overhead cables would allow and whilst extended and fixed to the roof of an electric train, it would make contact with electric catenary wires overhead to collect power for the electric locomotive.

Railcar – airbrake   1872

The railcar airbrake is a braking system used on railcars and is activated through the use of compressed air using exactly the same principle of braking used in modern locomotives today. George Westinghouse was the inventor of this breakthrough system and he was granted a patent in 1871.

 ### Railway - passenger   1825

The Stockton to Darlington became the world's first railway to carry passengers and freight. George Stephenson developed a keen interest as an engineer whilst working as a coal miner in North East England. His management team recognised his evolving skills and allowed him to experiment and build machines at his colliery.

When Stephenson first started work in the mining industry, coal-laden carts were pulled by horses to and from the mine and on to the nearby shipyards. Stephenson decided to use steam-powered engines to replace horses, leading to a succession of pioneering world firsts. The most significant of which was the world's first passenger railway between Stockton and Darlington.

 ### Railway - first inter-city   1830

The Liverpool to Manchester railway is the world's oldest 'inter-city' railway. The line has been constantly in use since its opening on 15th September 1830. Initially the link was designed to shift cargo away from the expensive and slow ship canal that ran between the port of Liverpool and the growing industrial hub of industry, Manchester. But within a year, such was its success, that the enterprising owners of the line decided to open it up to passengers and within months, it was carrying thousands of commuters.

 ### Railway - first steam passenger service   1830

The Canterbury and Whitstable railway became the first in the world to offer a regular steam passenger service.

Railway signals - semaphore   **1841**

Charles Hutton Gregory installed the world's first semaphore signals at New Cross Gate on the London to Croydon railway. This proved so successful that his technique was used worldwide on all railway systems until electric signals took their place completely.

 Railway signals - electric   **1893**

The world's first fully automated electric signaling system was installed on the Liverpool Overhead Railway.

 Railway tracks   **1603**

Huntingdon Beaumont opened the world's first above ground railway system from Wollaton mine to the city of Nottingham. Long before the advent of steam, electric or diesel power. Horses and ponies pulled the rolling stock on wooden tracks and sleepers.

 Railway tracks - cast Iron   **1767**

The world's first cast iron railway tracks were made and laid on the Coalbrookdale to Horsehay railway, England. The tracks were cast and made by a local iron and smelting works in Shropshire.

 Railway tracks - steel   **1857**

The world's first steel railway tracks were laid on the Midland Railway at Derby. Matthew Kirtley was the local railway superintendent and was instrumental in replacing the cast iron rails that were prone to fracturing and rapid erosion and pushed forward his idea of making the rails from the new 'Bessemer steel' using the world's first steel double-headed rail system, used globally today.

Refrigerator car   1867

Designed transport perishable freight by railroad, normally food produce, at specific chilled or sub-zero temperatures, the refrigerator car was a transformational invention. The original cars used either mechanical refrigeration units to chill the storage container, several crates of ice or a hybrid of both. The inventor of this ground breaking invention was J B Sutherland of Detroit, Michigan.

 ## Sleeping car   1839

A sleeper or sleeping car is a railroad passenger coach that provides sleeping accommodation for overnight journeys. The trailblazer of this new found travel luxury was the Cumberland Valley Railroad who launched the first sleeping car service in 1839, although it did not become commercially viable until the luxurious Pullman sleeping car was launched by George Pullman in 1857.

 ## Steam engine   1801

British engineer Richard Trevithick's invention would form the basis for all steam powered trains for years to come. In 1801 he tested a steam car, now known as 'the Puffing Billy'. It climbed the notorious Camborne hill in Cornwall successfully and Trevithick instantly became the first person to power an engine with a piston using high-pressure steam. This single invention was to revolutionize and transform the world of transport and commerce.

 ## Steam locomotive   1804

Richard Trevithick built the world's first steam railway locomotive. It ran on the Pen-y-Darren ironworks railway in Wales and carried the world's first railway passengers pulled by a steam locomotive.

Tilting train   1975

The APT (Advanced Passenger Train) was the world's first high speed tilting train, developed and built at British Rails Derby engineering works. It was tested successfully for the first time on the main rail network in 1975. It was capable of running at speeds of up to 155mph on the existing Victorian designed network, powered by a highly advanced control system that was well ahead of any other train design of its time. After a few minor problems with the tilt mechanism and a slight budget overrun, the state-owned engineering works was informed by the government that there was neither the political will nor funding to take the APT project to the next level. Tilting trains now run again in Britain, but are Italian built and ironically based on the same British design from the 1970's.

Train diesel   1976

The British class 43 locomotive known as the HST 125 high-speed train, is the world's fastest Diesel locomotive. Made by British Rails Crewe works between 1976 and 1983. The HST 125's top cruising speed is 125mph (200kmph). However, the world record for a Diesel locomotive was last set and recorded in the Intercity HST 125 Class 43 - 'City of Wakefield', running between Darlington and York on November 1st, 1987 when it smashed the previous HST record by running at 148.4mph (238.8kmph). The HS125 is still in service today in both Great Britain and Australia.

Train - world's fastest steam powered   1938

The super streamlined Mallard was and is still the world's fastest steam locomotive. Designed by Sir Nigel Gresley, who had already produced the world beating Flying Scotsman, the Mallard went on to achieve an as yet unbroken world record speed for a steam locomotive of 126mph (202kmph). This record was achieved by the 'Blue Streak' on July 3rd, 1938. It is unlikely that this record will ever be challenged.

🇺🇸 Trolley pole 1888

Sitting atop a sprung base of a trolley bus or light train to make electrical connection for motor power, the trolley pole started life, not as a cantilever or catenary, but as a sprung pole. This mechanism enabled the donor vehicle to maintain contact with the overhead powerlines, even when the bus or train ran over undulated surfaces. Never filing for a patent, Canadian John Joseph Wright is acknowledged to be the original creator, although a patent was granted in the same year to American Frank J Sprague of Richmond Virginia.

Underground railway - world's first 1863

The world's first underground railway, known as 'The Tube', opened as the Metropolitan Railway in London from Bishops Road, Paddington to Farringdon Street using steam engines to haul the gas lit carriages, that required air escapes to the street level to ventilate the tunnels of steam and toxic smoke from the trains. The unique underground railway was opened in 1863 and its first tunnels were excavated just below the street level using the 'cut and cap' method. Later extensions to the underground railway were dug much deeper through the clay and chalk terrain of London.

Due to the early use of steam engines, some grand Edwardian town houses were requisitioned and demolished, then rebuilt with just a solid facade to match the terraced houses either side. But the other side of the facade was completely open to the elements to expose the track below ground. This enabled engines to vent and let off steam and smoke. One such house still exists today, used only as a tunnel vent at 23/24 Leinster Gardens in Paddington, London. It is almost impossible to spot this house today where it still stands intact. It was not until 1890 that London's underground system was electrified and then the need for these dummy vent houses was no longer a requirement.

Road

 Airbag   **1952**

The air bag is now a standard feature on all automobiles to protect passengers in the event of a side or head on crash. The original airbag was invented by John W Hetrick in 1952. Following a car accident, he drew sketches of a container with compressed air that when a spring-loaded weight sensed the vehicle decelerate at a pre-determined rate, would activate a valve to instantly fill the bag and subsequently cushion and protect the passenger. He was granted a patent on this design on August 5th, 1952.

 Automatic transmission   **1904**

Invented and developed by 'The Sturtevant brothers' of Boston Massachusetts, automatic transmission is an automobile gearbox that changes gears dependent on torque as the vehicle is in motion. This concept freed the driver from manual gear shifts and this is the point at which modern auto gearboxes can trace their beginnings.

Automobile self-starter   **1911**

Commonly known as a car starter motor, the automobile self-starter is simply an electric motor that introduces a turning motion in an internal combustion engine (ICE). Prior to this device, all cars were started manually with a hand crank. The inventor of this breakthrough invention that is now used on all ICE automobiles today was Charles F Kettering.

 Bicycle - chain drive   **1873**

British bicycle designer and racer Henry John Lawson made the world's first chain drive safety bicycle. With huge advances in steel tube production, the new safety bicycle was made with a full steel tube frame, with a familiar design that is still used today. Metal was now strong enough to make small chain sprockets and the bike was now light enough for any person to power by pedaling a large cog via a chain to a smaller cog on the rear wheel. Thus, the first chain driven pedal bike was invented and put into mass production.

 Bicycle - lithium-ion powered lightweight electric bike   **2004**

The world's first commercially available electric bike, powered by the now ubiquitous lithium-ion battery technology, was first introduced by pioneering British based international e-bike designers and producers, Urban Mover. Prior to this technology, all electric bikes and cars relied on much heavier Sealed Lead Acid battery technology that weighed more than 6 times the weight of a similar specified Li-ion battery. The bike was called the Urban Cruiser and became popular across the world due to its lightweight propulsion system and compact lithium-ion battery pack. Now, lithium-ion battery technology powers the majority of e-bikes and electric cars worldwide.

 Bicycle - modern   **1870**

British Inventor and gardener James Starley developed and patented the Ariel lightweight bike. The first of the modern frame bikes put Great Britain and James Starley at the forefront of bicycle technology for more than 80 years, earning him the unofficial title of the 'father of the bicycle industry'. The main principles of the bicycle were to place the rider at the correct distance from the ground, placing the seat correctly in relation to the pedals and designing the handlebars so that the rider could produce the greatest force upon the pedals with the least amount of effort. The Ariel was the true global start of bicycle manufacturing in Britain. It was the first bicycle to utilize the 'tension-wheel' using pre-tensioned spokes. A design we all now take for granted and has not changed in more than 140 years.

Bicycle - safety type with pedal   1839
Foundry worker Kirkpatrick MacMillan from Glasgow devised a treadle powered wooden framed two-wheel bicycle and becomes the first mechanically driven bicycle in the world. MacMillan completed the treadle driven bike in 1839. It included wheels with wooden spokes and iron rims, a steerable front wheel and slightly larger rear wheel connected to the treadle pedals via iron rods.

Bicycle seat – padded   1892
Unlike a standard bicycle saddle at the time, the bicycle seat was specifically designed to support the riders back and buttocks. The padded bicycle seat was Invented by Arthur Lovett Garford from Elyria, Ohio and has since become ubiquitous around the world.

Bicycle tyre - pneumatic   1887
The world's first pneumatic bicycle tyre was made by John Dunlop in 1887. British vet John Dunlop first used his ground-breaking invention, the pneumatic tyre, on a bicycle. At last comfort and safety could be bought in the same package. This proposition was also getting cheaper with the advent of better mass production methods. Surprisingly, the chain was later eclipsed on these bikes with the development of a shaft drive to rid the bike of the dirty and often dangerous chain. But this proved very unpopular and only reappeared as recently as 2005 with a design very similar to the 19th century invention. The shaft drive is still unpopular due to the complexities of changing or replacing a rear wheel, compared to a simple chain driven bicycle.

Bus service - first inter-city **1831**
Entrepreneurs Sir Charles Dance & Goldworthy Gurney inaugurated the world's first steam powered inter-city bus service between Cheltenham and Gloucester.

 Bus - hybrid **1998**
British technology company BAE Systems designed and built the hybrid drive propulsion system that powers the world's largest fleet of hybrid buses in the world. Consisting of an ultra-low emission Diesel engine, powering a generator via a bank of high-performance lithium-ion batteries, that in turn drive electric motors in each wheel, using no mechanical transmission. Each system incorporates regenerative braking that feeds power back into the batteries, when braking is enabled. BAE's hybrid bus system is now powering low emission buses across the UK and USA.

 Bus - double decker **1909**
The world's first closed top double-decker buses were introduced by the Widnes Corporation in 1909. This effectively doubled the passenger capacity using a similar footprint as the original single deck buses that they replaced. Double Decker buses are now used globally.

 Car audio **1930**
A video or audio system, or combined audio/visual system fitted to a car is known by its generic term as car audio. Inventors Joe and Paul Galvin invented a basic car audio system in 1930 and introduced their first commercial car radio model which they named the Motorola model 5T71. Their first car radios sold for up to $130 USD, expensive even by todays standard.

Car - powered   1711

British inventor Christopher Holtum demonstrated a horseless carriage in 1711, 170 years before Karl Benz's horseless carriage. He gave demonstrations under the piazzas at Convent Garden in London and the car travelled at 6 miles an hour. This was the world's first powered car. Little is known about how the car was powered, but from the documented demonstrations in front of the public and press, it was clear no horses were involved.

 ### Car - electric   1884

Thomas Parker, a renowned electrical engineer and Victorian inventor, made and tested the world's first electric car and was photographed sitting in the vehicle in a picture revealed by Parker's great-grandson Graham Parker, an ex-BBC weatherman. Thomas Parker was also famous for electrifying the London Underground, creating overhead tramways in Birmingham and Liverpool and developing the smokeless fuel 'Coalite'. Sadly, Parker did not pursue his dream of electric cars, a technology overtaken at the time, by the development of the simpler and more powerful internal combustion engine. He died in December 1915.

 ### Car - world's fastest   1997

Thrust SSC is a British engineering speed car project that currently holds the world land speed record of 763mph, set on October 15th, 1997 in the Black Rock Desert in the state of Nevada, USA. The Thrust became the first car to officially break the sound barrier. Royal Air Force Pilot, Wing Commander Andy Green, piloted the Thrust SSC Supersonic Car to its successful world record breaking run.

 ### Car - world's fastest diesel   2006

At the Bonneville Salt Flats on August 23rd, 2006 the same man that was responsible for driving the world's fastest car, Thrust SSC, was behind the wheel of the world's fastest Diesel-powered car called the JCB Diesel Max. The car was designed and built by a team of engineers from the British earthmoving vehicle company JCB. Diesel Max reached a top speed of 350.092 mph and created a sensation in the world of Diesel power.

Car - world's fastest future car 2020c

Bloodhound SSC is the world's fastest supersonic car in development, with the aim of smashing the 1000mph barrier. This British project is based in Bristol and will be driven by RAF pilot Andy Green, already the holder of the current land speed record of 1997, in Thrust SSC. The Bloodhound SSC Super Sonic Car world record attempt will be made in the Kaksken Pan, South Africa, in a large flat part of the desert that is 12 miles long and 2 miles wide. Bloodhound SSC will be powered by a jet engine up to 300mph, then a rocket will take over to accelerate up to and beyond 1000mph. This takes 55 seconds to achieve, at which point air brakes and a parachute will slow the Bloodhound SSC down to 160mph, after which brakes made from steel rotors will then come into force to bring the car to its final halt. The car is 46 feet long and weighs in at a fraction over 7 tonnes, with most of the weight coming from the two engines that produce a thrust of more than 135,000 BHP.

 ## Car - first permanent 4-wheel drive car 1966

The Jensen interceptor was a British designed and engineered luxury sports car made by Jenson Motors, favored by the Princess royal; Princess Anne, who owned several during its production years in the mid-1960s to early 1970s. The Jenson interceptor was the world's first all-wheel, permanent four-wheel drive car of a non-all-terrain design and the first car to be fitted with anti-lock brakes as standard. It ceased production in 1971. This was a truly pioneering car of its time.

 ## Car - standard anti-Lock braking system 1966

British sports car, the Jensen interceptor was the world's first production car to be fitted with anti-lock brakes as standard. It used the Dunlop Maxaret mechanical system, previously the preserve of lorries, aircraft and racing cars.

🏴 Caravan   1885

Dr Gordon Stables was a British medical doctor who became the proud owner of the world's first caravan made by the Bristol Wagon Company. Built solely for holidaying, the caravan was called the Wanderer and is still owned by the Caravan Club in Broadway (GB) after a long-life travelling tens of thousands of miles. It is 30 foot long and weighs in at an incredible 2 tonnes. It is fitted out with an original china cabinet, bookcase and musical instruments in its opulent interior. Two horses pulled the Wanderer when it first toured the countryside in 1885 and has now been retired for posterity.

🇺🇸 Catalytic converter - three-way   1973

Providing an environment for a chemical reaction that converts dangerous combustion emissions into a reduced toxic output, the catalytic converter was co-invented by Carl D Keith and John J Mooney whilst employed by the Engelhard Corporation in 1973. Catalytic convertors are now used on most combustion engined vehicles.

🏴 Cats eyes - illuminated road demarcation   1934

British inventor Percy Shaw created the 'cats eye', now used globally as a self-illuminating and self-cleaning road demarcation unit. Percy Shaw was a road contractor from Yorkshire, who claimed the inspiration for his invention came when he drove home one night from his local bar and spotted the reflection of light in a cat's eye on the road. His invention was voted the greatest invention of the 20th Century in the Inventor Magazine.

🇺🇸 Convertible   1922

A convertible automobile is generally a sedan that has a canvas roof that can be converted into an open top vehicle, by folding back a canvas roof into the rear of the car behind the seats. Its inventor was Ben P Ellerbeck and fitted his first convertible roof on to a Hudson Coupe.

Child safety seat 1960

Often referred to as an infant safety seat, the child safety seat is a restraint system or child seat created to protect children from serious injury in the event of a collision. Its inventor, Leonard Rivkin from Denver, invented his first child safety seat for use with bucket seats. He was granted a patent on October 22nd, 1963.

Crash test dummy 1949

In 1949, Samuel W Alderson invented the crash test dummy after an extensive period of research and design, investigating real data in human test simulation, removing the guess work from automobile safety design. The crash test dummy is effectively a real-life full-scale human-like test device, built to the same average weight, proportions and flexibility of a human body.

Cruise control 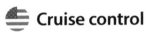 1945

An automatic driver set speed regulator of a motor vehicle; cruise control takes over the throttle after the selected speed is set to enable the car to maintain a constant speed. Cruise control was invented and developed by a visually disabled mechanical engineer called Ralph Teetor. The first vehicle to adopt Teetor's invention was the Chrysler Imperial in 1958.

Disc brakes  1902

British inventor Frederick William Lanchester created the disc brake to overcome the common problems encountered by conventional drum brakes, such as poor performance in the wet and uneven wear and tear leading to unreliability. His Disc brakes used brake pads fitted to a caliper that straddled either side of the rotating disc, fixed to the outside of the wheel hub. When the brakes were applied, both pads simultaneously squeezed each side of the disc to slow down and prevent the disc and wheel from turning.

Unfortunately, like so many great inventions, it was not until the 1950's after his death, that Lanchester's idea became commercialized. Disc brakes are now used as standard on almost all cars, motorbikes, buses, lorries and trains.

 ## Driving on the left   10AD

The majority of people are right handed and many years ago, when troops went into battle on a horse, they wanted to keep their strongest arm (right) free to hold a weapon. Therefore, if you position yourself on the left, then the enemy hopefully will be on your right and that's why to this day, Britain drives on the left. Strategically, it was always the best side to ride on in battle. Napoleon was left-handed and despite the majority of his army being right-handed, he demanded that they ride on the right. Rumors abound that this is the reason he lost the battle of Waterloo, because he insisted that all his troops use their left hand to hold their weapons to fight.

 ## Headrest   1921

A headrest is a head restraint that is attached to the top of a vehicle's seats directly behind the driver or passengers head. Cushioned for support, they are normally height adjustable and generally match the material and colour of the main seat. The inventor credited with designing the headrest was Californian Benjamin Katz, who was granted a patent in 1921.

 ## Lorry or truck   1870

The world's first practical lorry or truck designed for carrying freight was designed and built by Briton John Yule to carry ships boilers. The lorry was steam powered and was in use until the turn of the century.

 ## Lorry - articulated   1898

The world's first articulated lorry or truck was designed and built by Thornycroft Company Ltd. The lorry was powered by steam, driving a chain to power the front wheels.

Road

 Mini roundabout **1909**

The world's first circular junction (Roundabout) for road vehicles was built in Letchworth Garden City. During WW11, Frank Blackmore was a British traffic Engineer who developed the offside priority rule for roundabouts, through his work with the UK Road Research Laboratory. Blackmore also invented the Mini Roundabout, in use now in many countries throughout the world.

 Motorcycle – steam powered **1867**

A single track, two wheeled motor vehicle propelled by an engine, the first steam powered motorcycle was invented in 1867 by Sylvester Howard Roper and was known then as the Roper steam velocipede.

 Motorcycle - first petrol powered **1884**

Edward Butler was a British inventor who developed and patented the world's first motor powered tricycle. Butler called this innovative vehicle 'the Butler Petrol Cycle. Two Years before Karl Benz demonstrated his car in Germany, Butler revealed his three-wheeled petrol vehicle at the Stanley Cycle Show in London during 1884. Many consider quite rightly, that this was the world's first petrol powered car and also the world's first petrol powered three-wheeled vehicle.

 Muffler - silencer **1897**

Invented by Milton O Reeves, the muffler is a passive mechanical method of reducing the volume of noise discharged by an engine. The exhaust gases are forced out through the muffler on a conventional internal combustion engine. This pioneering invention was granted a patent in 1897. Ironically, the latest breed of electric cars are all being fitted with noise generators now, to warn people of their presence.

 One-way streets **1617**

The world's first official one-way streets were introduced and restricted to one-way traffic in the city of London. The trial of 17 streets was an effort by the civic authorities to avoid the growing congestion as London experienced huge population and traffic growth. The experiment was so successful that not only did all 17 trial streets remain one-way, but many more streets were also designated one-way in the following years of London's rapid expansion.

Parking meter   1935
Invented by Carl C Magee of Oklahoma City, Oklahoma, the parking meter is a machine that is primarily used to collect payment in return for the right to park a vehicle in a designated parking space for a specified length of time. Magee's invention was granted a patent on May 24th, 1938.

Powered steering   1926
A system on a powered vehicle to reduce steering effort whilst driving, powered steering was invented by Francis W Davis. The system provided assisted steering, particularly at low speed by using a supplementary power source normally either hydraulic or electrically supported.

Racing car circuit - first purpose built   1907
The world's first purpose built racing car track was built by British land owner, Hugh Locke-King. Work commenced construction in 1906 and the design remit was awarded to Colonel H.C.L Holden who came from the royal artillery. From Locke-Kings original plans for a road track, he was persuaded to revise the design so that participating cars could achieve the maximum achievable speed with the greatest safety levels built in.

The oval circuit was a total length of two and three-quarter miles and one hundred feet wide with two huge banked sections at each end of the oval, to ensure cars could maintain the greatest possible speed in relative safety when cornering. The concrete track was built in a record time of nine months and was one of the most remarkable achievements in civil construction of its time.

Radar gun — 1954
The radar gun is a small Doppler radar that is employed to detect the speed of objects, relying on the Doppler effect that is applied to a radar beam, measuring the speed of objects to which it is pointed. Primarily hand held but often vehicle mounted, the radar gun was invented by Bryce K Brown in 1954.

 ### Road surface marking — 1911
Invented by Edward N Hines and first used in Wayne County, Michigan in 1911, road surface markings describes any official local or national authority road markings to highlight official highway safety information, such as a highway centre line or road edge markings.

 ### Scooter - child's — 1912
The world's first commercial child's scooter or Ska-cycle (derived from skate-cycle) was patented by C.E. Richardson & Co Ltd, a company famous in its time for making Cyclecars in Sheffield. Before the First World War, they specialized in children's toys, including toy cars and model aircraft. The company later went on to design and produce more than 600 lightweight low power cars and exhibited each model at the Motor Car Show at Olympia in 1920.

 ### Spare wheel — 1904
British brothers Thomas and Walter Davies invented the world's first spare wheel after they saw a gap in the market following several reports of motorists being stranded after accidents or getting punctures. The 'Stepney Spare Wheel', named after the street where their ironmongery and bike shop was based, was an instant success when it was launched in 1904. The Davies brothers sold more than two thousand wheels a month across Europe and the USA at the peak of their business, before car companies decided to make them a standard fitment on all new cars.

Street lighting   1417

It is widely accepted that the Mayor of London, Sir Henry Barton, first ordered illumination of streets in public places by written public order. However, there is no further documentation to detail how this was introduced.

Street lighting - gas   1794

William Murdock was a British engineer and inventor who is acknowledged as the first person to design and install gas street lighting. Murdoch's idea came from his days working in Cornwall on a pump engine installation. Whilst relaxing by an open fire one night, he noticed that after placing a sprinkling of coal dust in his pipe, when placed in the fire, coal gas was produced, coming through the pipe stem and out of the mouthpiece. This was the point that he discovered gas from coal could be used as a lighting medium. He then set about installing a gas lighting system in his house in Redruth and this is recorded as being the world's first house to be lit by gas.

Street lighting   1807

Samuel Clegg was a British mining engineer and colleague of gas lighting pioneer, William Murdoch. Witnessing the potential of Murdoch's original gas lighting system, Clegg travelled back to London and set up his own business called the 'Gas Lighting & Coke Company'. On January 28th, 1807 his installed gas lighting system enabled Pall Mall in London, to become the first Street in the world lit at night by street lighting.

Taxi   1625

The world's first documented for-hire wheel-based taxi service related to London Hackney Carriages in 1625. The 'Hackney cabs' as they were known, consisted of an enclosed 2 or 4-seat coach with room at the back for luggage and privacy curtains for its passengers inside the upholstered compartment. The cabs were horse drawn and were generally available to hire from London inn Keepers. The world's first taxi rank was created in London on the Strand in 1635 and in the same year an act of Parliament was passed to make horse drawn carriages for hire legal in the City of London. The first taxi licenses were issued in 1662.

Traffic cone 1914

Invented by Charles P Rudabaker traffic cones are generally cone shaped and placed on sidewalks or roads as a temporary reroute or warning whilst road or construction work is in progress.

 Traffic lights - gas 1868

British Inventor John Peake-Knight installed traffic lights at the junction of Great George Street and Bridge Street near the Houses of parliament, London. The lights were operated by gas. Globally, traffic lights have changed the world

 Traffic lights - electric 1912

Also known as traffic signals, the traffic light is a signaling device generally situated on a road intersection to control traffic flow safely. Using a series of colors, normally red (stop) amber (get ready) and green (go), or often just red and green are used. Policeman, Lester Wire from Salt Lake City, Utah invented the first red-green electric powered traffic lights.

 Tow-truck 1916

Used to transport automobiles and other vehicles from one location to another, a tow truck operates primarily from a repair garage. The tow truck was invented by a garage engineer Ernest Holmes Senior of Chattanooga, Tennessee.

 Vehicle road tax 1921

Britain introduced the world's first national vehicle tax (VET vehicle excise tax) that applied to all 4-wheeled vehicles. This disc was to prove payment of the tax to use a vehicle on a public road. Ironically, on October 1st, 2014 road tax discs were abolished in Great Britain due to advances in electronic registration and camera enforcement.

GB & USA: The Mothers of Invention

 Wheel clamp   1953

Also known as the Denver boot, the wheel clamp is synonymous with preventing illegally parked vehicles, whose owners have their vehicles clamped. It is a very simple but effective concept that consists of a clamp that surrounds a vehicle wheel and locks in place to prevent the vehicle from being moved, or as a deterrent to stop thieves stealing vehicles. Its inventor was Frank Marugg of Denver, Colorado who was granted a patent on July 28th, 1958.

 Windshield wipers   1903

A bladed device employed to wipe rain and grime from a windshield of a road vehicle, the windshield wiper or windscreen wiper as it is known outside the USA, was invented by Mary Anderson. She described it as 'a window cleaning device for electric cars and other vehicles, operated by a lever from inside the vehicle'. Although later motorized, Andersons invention closely resembles the wipers found on early models for the following two decades. She was granted a patent on November 10th, 1903.

> "Have you ever noticed that anybody driving slower than you is an idiot, and anyone going faster than you is a maniac?"
>
> George Carlin - comedian and actor

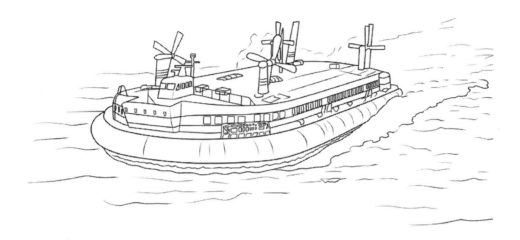

Sea

 Amphibious vehicle   **1805**

An amphibious vehicle is able to operate and be used on both land and water. Oliver Evans invented the 'Orukter Amphibolos' whose steam engine powered both wooden wheels and paddles and was intended for use as a multi-purpose means of transport. This was described as the USA's first land and water transporter.

 Chronometer - marine   **1761**

John Harrison invented the chronometer for accurate navigation at sea. Precise navigation at sea has always been extremely important, but until the development of the marine chronometer, it was almost impossible. It wasn't until 1714 that the British Government offered a huge cash incentive of £20,000 for somebody to solve this problem. British inventor John Harrison had spent his whole working life trying to solve this problem and eventually got his reward for developing the marine chronometer, worth more than £3 Million in today's money.

GB & USA: The Mothers of Invention

 ### Diving bell   1691

Edmund Halley was a medical doctor, astronomer, mathematician, physicist and meteorologist. He finalized designs for a diving bell that enabled the user to remain submerged for long periods of time. Halley created a window on the front face so that the diver could observe and explore freely. To replenish air, Halley's design included weighted barrels of air that were sent down at set intervals, allowing the diver to extend their underwater explorations. Halley is best known for calculating a comets orbit that was ultimately named after him, following verified proof and acceptance in the world of astronomy.

 ### Greenwich meridian   1884

The Greenwich Meridian is the line that divides the eastern Hemisphere with the west and is the official centre of world time. GMT is used to set every single timepiece in the world. GMT was officially adopted globally in 1884.

 ### Flat boat   1782

Used primarily for freight and passengers, a flat boat is a rectangular flat-bottomed vessel with square ends and generally used in calm inland waterways. Its inventor Jacob Yoder built a large prototype version of his design, which he loaded with bags of flour at Redstone Old Fort and transported it to New Orleans in May 1782.

 ### Hovercraft - amphibious   1953

Christopher Cockerell was a inventor from Cambridge who designed and built the world's first hovercraft called the SRN1. Cockerell owned a small boat building company and was looking at innovative ways to increase the speed of boats by reducing drag. He knew that Thorneycroft had experimented with a prototype 'air boat' where a huge top mounted engine powered a large fan, creating a huge uncontrolled downdraft, lifting the vessel slightly, but was hugely inefficient. Cockerell set about designing a more effective method of uplift by introducing rubber side skirts, containing the air and providing greater lift from a smaller motor and fan, to which he patented his design in 1955. It was built by Saunders-Roe and launched on 30th May 1959.

 Hydrofoil 1899

British naval architect and inventor John Isaac Thorneycroft founded the now famous Thorneycroft shipbuilding company. Thorneycroft developed and introduced many innovations over the following years, but the most significant was his Hydrofoil principal. While working on a method of hull lubrication by air, he experimented with underwater 'wings' to lift the boat and therefore reduce drag, enabling the boat to travel at higher speeds with less power required. This work developed into the first version of hydrofoils called 'stepped chine hulls' that Thorneycroft fitted and used successfully for naval motorboats in WWII.

Octant 1730

Also known as a reflecting quadrant, the octant is a navigational measuring instrument. By using mirrors to reflect the light path to the observer, it doubles the angle measured, allowing the device equipped with a one eighth arc to measure a quadrant or quarter circle. Although the octant was invented by John Hadley of Philadelphia, English mathematician and inventor Thomas Godfrey developed a similar instrument simultaneously, giving legitimate claim to the invention by both men.

Personal watercraft 1973

Commonly known as a wave runner or jet ski, the personal watercraft is a recreational vessel in which the rider sits or stands and steers and controls the speed using handlebar controls. Most variants use a combustion engine that drives a screw impeller that creates a jet thrust out of the back base of the craft. The inventor credited with this popular craft was Clayton Jaconson in 1973.

 Plimsol line 1870c

British politician Samuel Plimsoll from Bristol conceived and developed the Plimsoll line for ships. His invention consisted of a horizontal line on the hull of a ship to denote the maximum safe payload and therefore the safe free space available, depending on operational requirements and conditions. This simple mechanical method of ensuring payload safety in shipping is still used on ships globally.

GB & USA: The Mothers of Invention

 ### Ship-steam   1838
The SS Great Western started the world's first transatlantic paddle steamer service. This great craft was designed and built by British engineer, Isambard Kingdom Brunel.

 ### Ship - propeller driven   1838
The Archimedes was a British designed and manufactured ship and when launched in 1838 became the world's first screw propeller driven ship.

 ### Solar powered boat   1975
The world's first documented solar powered boat is a vessel built by British inventor Alan T Freeman and named 'solar craft'. Making its maiden voyage on 19th February 1975 on the Oxford Canal at Rugby. The 'Solar Craft' was fitted with ten Lucas solar modules and enabled the 2.5M long catamaran to travel at a lowly speed of 2.5mph.

 ### Steel hulled ship - with screw propeller   1843
British engineer and prolific inventor Sir Isambard Kingdom Brunel designed and built the first passenger liner in the world with a steel hull and screw propeller. The SS Great Britain was at the time the longest passenger ship in the world and built by Brunel for the Great Western Steam Ship Company for their service between Bristol and New York. She was the first iron-hulled liner to cross the Atlantic in 1845 and took 14 days for the crossing. After being rescued from the Falkland Isles where it was being used as a warehouse, the SS Great Britain is now in dry dock in Bristol fully restored and is an award-winning visitor attraction.

Stadimeter 1894

A type of optical range finding device, the stadimeter is used to assess the distance of an object of known height, by measuring the angle between the bottom and the top of the entity. Similar in many ways to a sextant in the way that it measures an angle between two objects using mirrors, although the stadimeter dials in the height of the known object prior to measurement. The stadimeter was invented by Rear Admiral Bradley Allen Fiske of the US Navy in 1894. Fiske was granted a patent on July 31st, 1894.

Conclusion

The metaphor of choice for 'progress' is usually a locomotive: an engine surging forward on a linear course towards a scintillating future. But the path of innovation and invention is almost never linear. Discovery happens on a landscape of competition, fits, starts, explosions, thrusts, failures and grand successes. For much of modern history, the central landscape of discovery was a collection of small Islands in the North Atlantic. For the past three hundred years, Britain and the USA have served as the innovation incubators for the world.

For so many of the technologies that we take for granted as conveniences of the modern world, their story begins either in Great Britain or the USA.

Long before the modern aviation pioneers were even born, the science of flight had been elucidated by British pioneers such as Sir George Cayley and Percy Pilcher. Although, it is acknowledged that the world's first manned, heavier than air powered and controlled flight was laid claim by the renowned American pioneers of powered flight, the Wright brothers.

Before Sir Alistair Pilkington invented float glass, the world was a far hazier place. Now, float glass, which allows for uniform thickness and flat surfaces, is the most widely used form of glass in the world.

Britain produced the world's first steam engines, passenger railways and the underground train network.

In medicine, British and American scientists have helped save many millions of lives with the development of antibiotics, vaccinations, and anesthetics. Equally important are the myriad of Anglo-American discoveries in basic biology that have helped lay the foundation for the medications of the future such as monoclonal antibodies that were first humanized by Greg Winter in 1986.

The spread of the English language, for better or worse, has meant that it is now the de-facto lingua franca in science, business, aviation and the Internet.

Conclusion

Britain ranks number one and American number two in global Nobel Prize winners per capita.

The scale of these discoveries and innovations raises one major question: how? How is it that such a small island nation and relatively new world country contributed so much to modernity?

The answer lies in the creation of a system of scientific discovery that is deeply rooted in education and a culture of curious engagement with the world, premised on the idea that human reason can uncover the secrets of the universe.

This system that is common in both countries, has produced a continuous supply of global innovators and leaders. It should not come as a surprise that the USA hosts the world number one university and Great Britain remains home to number two and three, both the oldest in the English-speaking world.

Leading British and American thinkers have done more than just invent. They created the system of rationality that allows for invention. Roger Bacon, an Oxonian is credited with developing the first formulation scientific method and may be the preeminent example. Bacon (1214–94) was promoting a theory of scientific methodology 700 years before the world's scientists finally accepted it as the dominant form of conducting research.

Throughout the last 200 years, Britain and America continued the tradition of nurturing brilliant minds at the forefront of science and its advancement.

Many British people led at the vanguard of the Scientific Revolution. A prime example is William Gilbert (1544–1603), who laid the foundations for the theory of magnetism and electricity.

Around the same time, William Harvey demonstrated that blood circulates through the human body.

Sir Isaac Newton advanced the law of universal gravitation and through his works, showed that scientific theory should be coupled with rigorous experimentation, which became the basis of modern science.

Britain stood at centre-stage in the Industrial Revolution that spanned the 17th and 19th centuries. Though it was British and American people and firms together that helped lead this world historical transformation and fundamentally changed how commercial activity takes place, how products are manufactured and how society is organized.

In the midst of these social changes, British scientists made earth-shattering discoveries of their own: Charles Darwin introduced his radical theory of evolution by natural selection in 1859, and Michael Faraday equipped later generations of inventors and physicists with the principles of electricity, magnetism, motors and generators and American Dr. Jonas Salk created a vaccine for polio in 1952 that has now contributed in almost eradicating the disease globally.

The scientific prowess of British and American inventors, philosophers and innovators continued right through the 20th century.

Alexander Fleming, the pharmacologist, discovered and developed antibiotics with his discovery of penicillin. British and American innovators could be found across virtually every field of inquiry: Ernest Rutherford, the father of nuclear physics, Nikola Tesla, the father of modern technology, John Logie Baird, the father of television, and Alan Turing, the father of theoretical computer science and artificial intelligence.

There has been no deceleration in the rate of Anglo-American innovation, even in the fully globalized 21st century. Both countries are still world leaders in modern medicine, engineering, science, music and literature.

Given the full weight of British and American achievement in science and technology, it would be reasonable to imagine that humanistic and artistic pursuits would be less emphasized. But nothing could be further from the truth.

Conclusion

British and American poets and writers have left indelible marks on the world throughout the centuries: William Shakespeare, Mark Twain, Virginia Wolf, Byron, Arthur Miller, George Orwell, F Scott Fitzgerald, Shelley, Henry James, Charles Dickens, Stephen King, George Eliot, and JK Rawlings. The international impact of British and American musicians and composers such as The Rolling Stones, Elvis Presley, The Kinks, The Eagles, The Beatles, Michael Jackson, Queen, Billy Joel, Pink Floyd, Holst and Elgar, cannot be denied. The list is endless.

The story of British and American creative success is a great topic for historians. But it also holds great lessons for the future. It is a lesson in how to construct an innovative cultural and social system that sustains high quality output over centuries. Britain and America are indeed the global "mothers of invention," and we would do well to better understand and preserve those forces that have allowed both great countries to contribute so much to humankind.

About the author

Moulded and motivated by severe hardship as a young boy, Keith grew up in a 'two up, two down' Victorian terrace in the heart of Birmingham, UK until the age of 4, with only an outside toilet and a portable zinc bath for washing next to the coal fire. But this abject poverty did not prevent him from developing an insatiable love for learning and an avid interest in all things historic; a passion encouraged by the many libraries, museums, and historical sites in his local community.

An entrepreneur by nature, Keith undertook his first business venture at the tender age of 11, buying jug and basin sets and wind-up gramophones to refurbish before selling them on to antique and second-hand dealers in Dover and Oxford. Then, as a teen when he left school in surfing town, Newquay, his father wanted him to be a postman - a practical job with security! Instead, Keith pursued his dream of becoming an electrical engineer. He later attained a master's degree in Business Administration and is currently working towards his PhD in zero carbon engineering technologies, as he simultaneously creates and collaborates on revolutionary advancements within the green-tech, mixed energy and zero carbon sectors.

Keith has not only worked for himself but also directly for a number of large, multi-national corporations, including Chief Technical Officer for a global American data centre company, Head of infrastructure deployment for Orange, and CEO and non-executive director in several green-tech enterprises. He is currently Consulting for numerous global energy storage solution and EV charging companies. Additionally, he is a Fellow of the Chartered Management Institute (FCMI) and Member of the Institution of Engineering and Technology (MIET).

An enthusiastic historian, Keith's first published book was about Britain's long record of invention and innovation, explaining how and why this helped shape the world as we know it. His unwavering thirst for knowledge has led him to travel extensively across the USA, Europe, and Asia, particularly Japan and China. A prolific cook, in his spare time he enjoys creating new garlic free and gluten free recipes, spending time anywhere

About the author

near the ocean, and continuing his lifelong quest for knowledge with the firm belief that continuous learning is the key to longevity - along with great food of course! Now with four grown-up daughters, he lives in the Cotswolds, but still finds time to manage several online companies as well as writing and publishing a wide variety of non-fiction works, from cookbooks to Industrial history research and more recently, zero carbon technical publications.

Chronological index by year

10AD Driving on the left	276
1170 America - discovery and settlement	57
1250 Magnifying glass	227
1299 Bowls - the sport	240
1386 Clock - Oldest continually working clock	28
1400c Golf	242
1400c Rounders	245
1417 Street lighting	280
1449 First British patent	13
1475 First English printed book	160
1483 Oil - the barrel measurement	134
1509 Cornish pasty	96
1534 Cambridge University Press	157
1557 Equal sign	133
1564 born Shakespeare	165
1578 Submarine	144
1583 The commonwealth	61
1589 Knitting machine	89
1596 Toilet - flushing	116
1600 Magnetic earth	226
1603 Railway tracks	264
1605 Cipher - Bacons cipher	43
1614 Logarithms	133
1617 One-way streets	277
1622 Slide rule	134

Chronological index – by year

Year	Entry	Page
1625	Taxi	280
1627	Water desalination	175
1628	Blood circulation - discovery of	205
1630	Telescopic sight	84
1638	Micrometer	79
1650	Coffee houses - coffee shops	96
1652	High pressure bottle - Champagne bottle design	95
1655	Spectrum heterogeneity of light	232
1660	Balance spring	28
1662	Champagne or sparkling wine	95
1663	Red blood cells	182
1664	Binary stars	176
1665	Compound microscope - 30 X magnification	193
1665	Gravity - the law of	221
1668	Metric system	133
1668	Telescope - reflecting	234
1669	Calculus	132
1676	Universal joint - commercial design	86
1680	Postage stamps	34
1680	Sign language	35
1684	Fingerprinting	42
1686	Jet propulsion - theory	222
1687	Motion - the law of	228
1691	Diving bell	284
1698	Steam engine - first steam engine	83
1700	Industrial revolution starts	119

1700c Mass production	22
1701 Seed drill	18
1705 Halley's comet	177
1705 Steam engine - atmospheric	83
1709 Blast furnace - world's first coke fired	123
1711 Car - powered	272
1711 Tuning fork	150
1714 Typewriter	168
1717 Swim fins - flippers	248
1718 Machine gun	138
1720 English horn	147
1720c Maritime clock - marine chronometer	172
1723 Water sprinkler	68
1725 Stereotyping	166
1727 Bridge - railway	72
1727 Surgical forceps	211
1728 Speed of light	233
1730 Octant	285
1730 Spectacles - modern side arm type	175
1733 Blood pressure - measurement	205
1733 Pram - perambulator or baby carriage	111
1747 Controlled clinical trials	194
1748 Friction match	107
1748 Fridge	107
1749 Lightning rod	225
1750 Horse racing - thoroughbred	243

Chronological index – by year

1752 Sunglasses	175
1755 Baseball	239
1755 Dictionary	158
1761 Chronometer - marine	283
1762 Sandwich	99
1763 Aspirin - discovery of the active ingredient	204
1764 Spinning jenny	92
1767 Railway tracks - cast iron	263
1768 Spinning frame	92
1769 New Zealand - discovery and settlement	65
1700c Toothbrush	117
1770 Australia - formation	59
1770 Caterpillar tracks - patent	73
1770 Eraser	106
1772 Figure skating	242
1772 Fizzy drinks - carbonated drinks	97
1772 Nitrous oxide - discovery of	187
1774 Concrete - quick drying	124
1774 Cricket - first rules	241
1774 Lifeboat - world's first	67
1774 Oxygen - discovery	188
1775 born Jane Austen	162
1775 Toilet - S-trap	116
1776 Hydrogen - discovery	187
1776 Lifeboat station	67
1776 Threshing machine	18

GB & USA: The Mothers of Invention

1776 Vaccination	213
1778 Lock - tumbler	78
1779 Bridge - iron	71
1779 Spinning mule	93
1780 Steel pen	114
1781 Uranus - discovery	180
1782 Flat boat	284
1782 Steam engine - modern condensing	83
1784 Bifocals	170
1784 Lock - tumbler - high security	79
1785 Loom - powered	90
1788 born George Byron	160
1788 Flyball governor	75
1790 First US patent issued	14
1790 Sewing machine	91
1790 Shoelaces	114
1791 Gas turbine	76
1794 Ball bearing	70
1784 Colour blindness - discovery	193
1794 Street lighting - gas	280
1795 Corkscrew	104
1795 Hydraulic press	126
1795 Wheel cypher	44
1796 Vaccination	213
1797 Banknotes - first public issue	103
1799 Anesthetics - nitrous oxide	202

Chronological index – by year

1800c Bank notes - world's largest manufacturer	71
1800c Billiards	240
1800c Hockey	243
1800 Light bulb - electric incandescent	224
1800c Manufacturing revolution starts	121
1800c Polo - sport	244
1800c Seat belt	81
1800c Wellington boots	87
1801 Bridge - suspension	72
1801 Dalton's law - of partial pressures	217
1801 Steam engine	265
1802 Clouds - categorisation of	171
1804 Rocket solid fuel	141
1804 Steam locomotive	265
1805 Amphibious vehicle	283
1805 Atomic theory	170
1805 Refrigerator	112
1806 Coffee percolator	104
1807 Chemical electrolysis	185
1807 Potassium - first isolation	188
1807 Sodium - first isolation	189
1807 Street lighting	280
1808 Aluminum - discovery of	122
1808 Boron - the isolation of	185
1808 Tension spoked wheel	84
1809 born Edgar Allan Poe	159

1810 Tin can	100
1813 Circular saw	74
1814 born Charles Dickens	158
1815 D.H. Lawrence	158
1815 Safety lamp	80
1816 Charlotte Brontë	158
1816 Fire extinguisher - modern portable	67
1817 Parkinson's disease - discovery of	200
1818 Blood transfusions - human to human	205
1818 born Emily Jane Brontë	159
1818 Ice hockey	243
1819 born George Elliot	160
1819 Hay fever - discovery	197
1819 born Herman Melville	161
1820 born Anne Brontë	157
1820 Computer - world's first	49
1820 Gin & tonic	98
1820 Macadam - road surface	127
1820 Squash	247
1821 Electric transformer	218
1823 Electromagnet	219
1823 Lancet - first medical journal	163
1823 Lift or elevator	78
1823 Mackintosh Waterproof Coat	90
1823 Rugby union	245
1824 Cement	124

Year	Entry	Page
1824	Fire brigade	66
1825	Benzene - first isolation	185
1825	Railway - passenger	263
1826	Stove - gas	114
1827	Friction match	107
1827	Lawnmower	109
1828	Mechanical reaping machine	18
1828	Thermosiphon - conductive heating system	85
1829	Concertina	146
1829	Police - world's oldest police force	66
1830	Adjustable spanner or wrench	69
1830	Elastic fabric	88
1830	born Emily Dickinson	159
1830	Railway - first intercity	263
1830	Railway - first steam passenger service	263
1830	Thermostat	85
1831	Bus service - first inter-city	271
1831	Chloroform - discovery	186
1831	Electric generator	218
1831	Electric motor	218
1831	Flanged T-rail	261
1831	Multiple coil magnet	228
1831	Safety fuse	129
1832	Dental chair	196
1832	born Thomas Hardy	167
1833	Sandpaper	129

1833 Sewing machine - lockstitch	113
1834 Combine harvester	16
1834 Mail order	22
1835 Photography - practical	155
1835 Relay - electrical	231
1835 Steam shovel	130
1835 born Mark Twain	163
1836 Arsenic test	40
1836 Circuit breaker	216
1836 Faraday cage	219
1836 Morse code	33
1836 Screw propeller	81
1837 Shorthand	166
1837 Telegraph - electric	35
1838 Internal combustion engine	77
1838 Ship - steam	286
1838 Ship - propeller driven	286
1838 Stereoscope - 3D	155
1838 Typhoid - discovery and prevention	211
1839 Bicycle - safety type with pedals	2
1839 Sleeping car - rail	265
1839 Steam hammer	130
1839 Vulcanized rubber	87
1840 Boolean algebra - digital logic design	132
1840 Electro plating	125
1840 Water and sewerage systems	131

Year	Entry	Page
1841	Clock - first electric	29
1841	Hypnosis	208
1841	Railway signals - semaphore	264
1841	Screw threads - standardization	81
1842	Fuel cell	220
1842	Grain elevator	17
1843	Computer programme - first programmer	49
1843	Conservation of energy - law of	216
1843	Fax machine	51
1843	born Henry James	161
1843	Ice cream maker - hand cranked	98
1843	Joule - thermodynamics	223
1843	Multiple effect evaporator	187
1843	Rotary printing press	129
1843	Steel hulled ship - with screw propeller	286
1845	Baseball - codified rules	239
1845	Diamagnetism	217
1845	Rubber band	80
1845	Tyre - Pneumatic	86
1846	Telegraph - printing	36
1846	Traverse shuttle	93
1847	Anesthetics - chloroform	203
1847	Chocolate bar	95
1847	Gas mask	137
1847	Hydraulic crane	76
1847	Neptune - discovery of	178

1848 Aircraft - first powered flight	251
1848 Bridge - box girder	71
1848 Chain stores - worlds first	22
1848 Zero	236
1849 Aircraft - first manned heavier than air flight	252
1849 Jack hammer	77
1849 Safety pin	113
1850 Bessemer converter - mass production of steel	123
1850 Mercerized cotton	90
1850 Robert Louis Stevenson	164
1850c Snooker - origin	246
1850 Transatlantic cable	37
1851 Oil - first oil tycoon	127
1851 Oil refinery	128
1852 Elevator brake	74
1852 Thesaurus	167
1853 Aircraft - first manned, controlled flight	252
1853 Aquarium - design and construction	70
1853 Burglar alarm	73
1853 Hypodermic syringe	199
1853 Potato chips	99
1854 Breast pump	191
1854 Cholera - discovery	206
1854 Cipher - Playfair Cipher	43
1854 Concrete - reinforced	125
1854 Pie chart	134

Chronological index – by year

1855 Plastic	172
1856 Condensed milk	96
1856 Dinosaur - nomenclature	181
1856 Radio first transmission and receiving	230
1856 Venn diagram	135
1857 Female doctor	207
1857 Postcodes or Zip codes	34
1857 Railway tracks - steel	264
1857 Toilet paper - mass produced	117
1858 Epilepsy - treatment	196
1858 Gray's anatomy	207
1858 Ironing board	108
1858 Pencil eraser	111
1859 born Arthur Conan Doyle	157
1859 Electric stove - cooker	106
1859 Escalator	125
1859 Evolution - by natural selection	182
1860 born J.M. Barry	162
1860 Fish and chips	97
1860 Linoleum	127
1860 Vacuum cleaner - manual	117
1861 Thallium - discovery	190
1861 Twist drill bit	85
1863 Breakfast cereal	94
1863 Football - FA	241
1863 Underground railway - world's first	267

1864 Microwave	227
1864 Spar torpedo	143
1865 Boxing - Queensbury rules	240
1865 Antiseptic surgery	203
1865 born Rudyard Kipling	164
1865 Periodic table	188
1865 Printing press - rotary roll feed	163
1865 Sewage systems	121
1866 born Beatrix Potter	157
1866 born H.G. Wells	161
1866 Torpedo	144
1866 Urinal	131
1867 Barbed wire	122
1867 Motorcycle - steam powered	277
1867 Paper clip	111
1867 Refrigerator car	265
1867 Ticker tape	37
1868 Ailerons	251
1868 Helium - first observation on the sun's surface	186
1868 Standard deviation - probability	135
1868 Traffic lights - gas	281
1868 Umbrella	117
1869 American football	238
1869 Clothes hanger	104
1869 Pipe wrench	128
1870 Bathtub - cast enamel	103

Chronological index – by year

1870 Bicycle - modern	269
1870 Can opener	103
1870 Lorry or truck	276
1870c Plimsol line	285
1871 Wind tunnel	259
1872 Diner	97
1872 Railcar brake	263
1873 Badminton	238
1873 Bicycle - chain drive	269
1873 Jeans	107
1873 Knuckle coupler	261
1873 Light - nature of electromagnetic theory	225
1873 Radio - theory of electromagnetism	231
1874 Electric cooking utensil	105
1874 Fire sprinkler - automated	74
1874 Forstner bit	75
1874 born Gustav Holst	148
1874 Qwerty keyboard	55
1874 Tennis	248
1875 Dental drill	196
1876 Synthesizer	150
1876 Tattoo machine	84
1876 Telephone	36
1877 Bridge - concrete	71
1877 Phonograph	33
1878 Bolometer	215

1878 Microphone	33
1879 Cash register	21
1879 Cathode ray tube	171
1879 Photographic plate	155
1880 Hydroelectricity	221
1880 Light bulb	224
1880c Table tennis or ping-pong	248
1881 Iron - electric	108
1881 Metal detector	79
1881 Peristaltic pump	200
1882 Electric fan	105
1882 born Virginia Woolf	168
1883 Fibre - man made	89
1883 Whistles	39
1884 Car - electric	272
1884 Dissolvable pill	195
1884 Greenwich meridian	284
1884 Light switch - quick break technology	224
1884 Machine gun - automatic	139
1884 Motorcycle - petrol powered	277
1884 Multiple-wheel steam turbine	79
1884 Skyscraper	82
1884 Steam turbine	83
1885 Caravan	274
1885 Fuel dispenser	75
1885 Photographic film	155

Chronological index – by year

1886	Telephone directory	166
1887	Bicycle tyre - pneumatic	128
1887	Clinical thermometer	194
1887	Comptometer	133
1887	Gramophone record	32
1887	Pneumatic tires - air inflated	270
1887	Slot machine	114
1887	Softball	247
1887	Theatre organ	150
1888	Ballpoint pen	102
1888	Drinking straw	105
1888	Induction motor	221
1888	Revolving door	80
1888	Solar power - first solar cell patent	232
1888	Spark plug - world's first	82
1888	Stepping switch	233
1888	Telautograph	35
1888	Touch typing	167
1888	Trolley pole	267
1888	born T S Eliot	167
1889	Cinematography - creator	153
1890	born Agatha Christie	156
1890	Babcock test	184
1890	Diesel engine - pre-Diesel patent	74
1890c	Netball	244
1890	Smoke detector	82

1890 Tabulating machine	135
1891 Basketball	239
1891 Dow process	186
1891 Kettle	108
1891 Rotary dial - telephone	34
1891 Synthetic rubber - isoprene	175
1891 Tesla coil	234
1892 Bottle cap - crown cork	70
1892 Dimmer switch	217
1892 Gasoline engine - with forward and reverse	17
1892 Newspaper - first printed in Braille	163
1892 Rayon - viscose	91
1892 Tractor - powered by gasoline	18
1892 Vacuum flask	118
1893 Lawnmower - steam powered	109
1893 Pinking shears	91
1893 Railway signals - electric	264
1893 Spectroheliograph	179
1893 Toaster - electric	115
1893 Zipper or zip	118
1894 Argon - discovery and proof	184
1894 Stadimeter	287
1895 Rayon - acetate	91
1895 Rugby league	245
1895 Volleyball	249
1896 Darts - traditional pub game	241

Chronological index – by year

1896 born F Scott Fitzgerald	160
1896 Seismograph	174
1896 Typhoid - first vaccine	212
1897 Cotton Candy	97
1897 Electron	219
1897 born Enid Blyton	159
1897 Malaria	209
1897 Muffler - silencer	277
1897 Ship - steam turbine	142
1898 Lorry - articulated	276
1898 Meccano	110
1898 Polyethylene	173
1898 Remote control	112
1898 Semi-automatic shotgun	142
1899 Aircraft - first powered controlled flight	253
1899 born Earnest Hemmingway	158
1953 Hydrofoil	285
1899 Mousetrap	110
1899 Plasticine	173
1899 Silicone	174
1900 Cartoon - first moving cartoon film	152
1900 Nickel zinc battery	229
1901 Assembly line - moving	21
1901 Mercury vapor lamp	227
1901 Radio direction finder	230
1901 Safety razor	113

1901 Tarmac	131
1901 Vacuum cleaner - powered	118
1902 Air conditioning	70
1902 Disc brakes	275
1902 born John Steinbeck	161
1902 Hearing aid - portable	198
1902 Periscope - collapsible	139
1902 Tea maker - automatic	115
1902 Teddy bear	115
1903 Aircraft - first manned powered flight	252
1903 Baler - round	16
1903 Caterpillar tracks - first commercial use	74
1903 born George Orwell (Eric Arthur Blair)	160
1903 Hormones - discovery	182
1903 Tea Bag	100
1903 Windshield wipers	282
1904 Automatic transmission	268
1904 Diode - rectifier tube radio Valve	51
1904 Pantograph	262
1904 Wendy - first use of the name	168
1904 Spare wheel	279
1905 Rolex	24
1906 ARG - electrically powered analogue computer	46
1906 Kinemacolour - colour movie process	154
1906 Dreadnought - first turbine powered battleship	137
1906 Flushometer	106

Chronological index – by year

1906	Typesetting	168
1907	LED - theory	223
1907	Racing car circuit - first purpose built	278
1908	born Ian Fleming	162
1908	Television - cathode ray tube	36
1909	Bus - double decker	271
1909	Mini roundabout	277
1909	Paper shredder	111
1910	Propane	189
1911	Atom - discovery of	169
1911	Automobile self-starter	269
1911	Road surface marking	279
1911	born Tennessee Williams	166
1911	Wing warping	260
1912	Autopilot	256
1912	Braggs law	215
1912	Electric blanket	105
1912	Isotopes - first concept of	222
1912	Scooter - child's	279
1912	SONAR	142
1912	Traffic lights - electric	281
1913	Formica	106
1913	Shaver - electric	113
1913	Stainless steel	130
1914	Acetylcholine	202
1914	Depth charge	137

1914 Fighter aircraft Vickers F.B.5.	138
1914 Tank - military	143
1914 Traffic cone	281
1915 born Arthur Miller	157
1915 Zener diode	235
1916 Condenser microphone	146
1916 born Roald Dahl	164
1916 Hamburger	98
1916 Light switch - toggle	225
1916 Supermarket	24
1916 Tow truck	281
1917 Aircraft - landing on a moving ship	253
1917 Stream cypher	44
1917 Vitamin A	213
1918 Aircraft carrier - HMS Argus	136
1918 Airforce - world's first	256
1918 Crystal oscillator	217
1918 Hydraulic brake	76
1918 RAM - Random Access Memory	55
1918 Torque wrench	85
1919 International daily air passenger service	257
1919 Mass spectrograph	227
1919 Silica gel	189
1919 Toaster - pop up	116
1921 Blood transfusion service	206
1921 Flow chart	133

Chronological index – by year

1921 Headrest - car	276
1921 Polygraph	42
1921 Vehicle road tax	281
1922 born Kingsley Amis	161
1922 Audiometer	215
1922 Convertible - car	274
1922 Radial arm saw	80
1922 Water skiing	249
1923 Bulldozer	73
1923 Instant camera	153
1923 born Joseph Heller	161
1923 Wristwatch - first automatic	29
1924 Cheeseburger	95
1924 Earth inductor compass	218
1924 Locking pliers	78
1924 Moviola	154
1924 Radio altimeter	257
1925 Automatic volume control	215
1925 Masking tape	109
1925 Television	37
1926 Powered steering	278
1926 Rocket - liquid fuel	141
1927 Bread slicer	94
1927 Jukebox	149
1927 Pressure washer	111
1927 Clock - quartz	29

1928 Iron lung	199
1928 Penicillin	210
1928 Razor - electric	112
1928 Television - colour	37
1928 Television - transatlantic transmission	37
1928 Video recorder	38
1929 Cyclotron	217
1929 Frozen food	98
1930 Car audio	271
1930 Jet engine	77
1930 Pluto - discovery of	178
1930 Thermistor	234
1930 Tiltrotor	258
1931 Cosmic radio waves	216
1931 Geological timescale	171
1931 Heavy hydrogen	186
1931 Stereo sound	35
1931 Strobe light	233
1931 born Toni Morrison	167
1932 Micro switch	228
1932 Neutron	229
1932 Radio telescope	179
1932 Splitting the atom	174
1933 Camera - multiplane	152
1933 Frequency modulation - FM	32
1933 Heavy water	186

Chronological index – by year

1933 Polyvinylidene chloride	188
1934 Cats eyes - illuminated road demarcation	274
1934 Trampoline	248
1935 Aerial refueling	250
1935 Parking meter	278
1935 PH meter	189
1935 RADAR - radio locator	140
1935 Richter magnitude scale	231
1935 Surfboard fin	247
1935 Ultra-violet - black light	235
1936 Chair lift -ski	240
1936 Fluorescent lamp - compact	75
1936 Guitar - bass	148
1936 Hay baling - automatic	17
1936 Philips head screw	80
1936 Programming languages	54
1936 Reed switch	231
1936 Stock car racing	247
1936 Strain gauge	233
1936 Surfboard fin	247
1936 Vitamin E	213
1937 Emergency calls - 999	31
1937 Niacin - Vitamin B	209
1937 Photosensitive glass	128
1937 Shopping cart	130
1937 Sunglasses - polarized	175

1938 Animal echolocation	181
1938 Fiberglass	125
1938 Nylon	187
1938 Soft serve ice cream	100
1938 Teflon	201
1938 Train - world's fastest steam powered	266
1938 Xerography	190
1939 ATM - concept	46
1939 RADAR - HS2 airborne radar	141
1940 Ejector seat	256
1940 Fluxgate magnetometer	219
1940 Plutonium	188
1941 Acrylic fiber	88
1941 Deodorant	104
1941 Guitar - solid body electric	148
1941 Polyester	91
1941 Special forces - SAS and SBS	142
1942 Bazooka	136
1942 Head-up display	257
1943 Computer - electronic programmable	50
1943 Magnetic proximity fuze	139
1945 Cruise control	275
1945 Microwave oven	110
1945 Penicillin - mass production	210
1945 Satellites - geostationary	173
1946 Credit card	22

Year	Entry	Page
1946	Diaper	105
1946	Formula 1 motor racing	242
1946	Space observatory	179
1947	Acrylic paint	69
1947	Defibrillator	195
1947	born Stephen King	166
1947	Supersonic aircraft	258
1947	Transistor	234
1948	Aircraft - world's first turboprop airliner	254
1948	Cable television	30
1948	Computer First stored programme	49
1948	Hair spray	107
1948	NHS - national health service	200
1948	Paralympics	244
1949	Atomic clock	214
1949	Aerosol paint	69
1949	Biometrics - Iris recognition	47
1949	Cataracts	192
1949	Crash test dummy	275
1949	Jet airliner - world's first	257
1949	Jet bomber	138
1949	Radiocarbon dating	230
1949	Soft whipped ice cream	99
1950	Smoking - lung cancer link	211
1950	Teleprompter	36
1950	Walmart	25

1951 Anesthetics - halothane	203
1951 Correction fluid	104
1952 Airbag	268
1952 Artificial heart	191
1952 Barcode	21
1952 Float glass	126
1952 Integrated circuit or microchip	222
1952 Polio vaccine	210
1952 Rapid eye movement	182
1952 born John Williams	149
1953 DNA - discovery	207
1953 Elvis Presley - 1953 to 1977	147
1953 Heart-lung machine	198
1953 Hovercraft - amphibious	284
1953 Voltmeter - digital	235
1953 WD-40	87
1953 Wheel clamp	282
1954 Acoustic suspension speaker	30
1954 Aircraft - world's first VTOL	254
1954 Fibre optics	31
1954 Radar gun	279
1955 Clock - first atomic	29
1955 Hard disk drive - mainframes	52
1955 Kettle - automatic	108
1955 McDonalds	23
1955 Nuclear submarine	139

1956 Industrial robot	77
1956 Nuclear power - generation station	229
1956 Video tape	38
1956 Ultrasound imaging	212
1957 Air-bubble packing	101
1957 Gamma camera	196
1957 Laser	223
1958 Carbon fiber	73
1958 Doppler fetal monitor	196
1958 Integrated circuit	222
1958 Sailboard and sail boarding	246
1959 Fusor	220
1959 Spandex	92
1959 Weather satellite	180
1960 Artificial turf	238
1960 The Beatles - 1960 to 1970	145
1960 Child safety seat	275
1960 Combined oral contraceptive pill	193
1960c Stun grenades	143
1961 Hip replacement	199
1961 Ibuprofen	208
1962 Cloning - theory	192
1962 Communication satellite	177
1962 Glucose meter	197
1962 Jet injector	199
1962 LED Light emitting diode	224

1962 Razor - stainless steel long-life blades	112
1963 Basic - computer language	47
1963 Carbon fibre	124
1963 Geosynchronous satellite	177
1964 Argon laser	214
1964 Beta blockers - propranolol	205
1964 8-track cartridge	147
1964 Hepatitis B virus - discovery	208
1964 Moog synthesizer	150
1964 Plasma display	34
1965 Aircraft - first commercial VTOL	254
1965 Baby buggies - strollers	101
1965 Chemical laser	216
1965 Cordless telephone	31
1965 Defibrillator - portable	201
1965 Hypertext	52
1965 Kevlar	89
1965 Packet switching	54
1965 born J.K. Rowling	162
1965 Snowboarding	246
1965 Space pen	232
1965 Yesterday - The Beatles	151
1966 ATM (cash machine) PIN	45
1966 ATM (Cash machine)	45
1966 Car - first permanent 4-wheel drive car	273
1966 Car - standard anti-Lock braking system	273

Chronological index – by year

Year	Entry	Page
1966	Mini skirt	24
1967	Back-pack - internal frame	102
1967	Pulsars - discovery	178
1968	Contraceptive pill	194
1968	Racquetball	245
1968	Virtual reality	55
1968	Workbench - portable	87
1969	Aircraft - first commercial supersonic airliner	255
1969	Laser printer	163
1969	Lunar module	177
1969	Taser	67
1969	Wide body aircraft	259
1970	Bar codes - commercial application	47
1970	Compact disc	146
1970	Glastonbury - music festival	147
1970	LCD Liquid Crystal Display	225
1970	Wireless local area network	56
1971	CAT or CT scanner	192
1971	Email	31
1971	Microprocessor	54
1972	Calculator - pocket	48
1972	Global positioning system - GPS	32
1972	Magnetic resonance imaging (MRI)	199
1972	Pet scanner	200
1973	Catalytic converter	274
1973	E-paper	159

1973 MRI Magnetic Resonance Imaging	209
1973 Mobile phone - cellular	33
1973 Personal watercraft	285
1973 RSA cipher	44
1973 Voice Mail	38
1974 Ball barrow	102
1974 Black holes	176
1974 Heimlich maneuver	198
1974 Scanning acoustic microscope	189
1974 Universal Product Code - UPC	24
1975 Digital camera	153
1975 Ethernet	51
1975 Microsoft	23
1975 Solar powered boat	286
1975 Tilting train	266
1976 Gore-Tex	89
1976 Hepatitis B vaccine	208
1976 Lithium-ion battery	226
1976 Swivel chair	115
1976 Train - diesel	266
1977 Apple	20
1977 Test tube baby	183
1978 ARM - universal mobile device processor	47
1978 IVF	209
1979 Computer - laptop	49
1979 iPod® Origins	53

Chronological index – by year

Year	Entry	Page
1979	Winglets - aircraft	260
1980	MRI machine	210
1981	Computer - portable	50
1981	Graphical user interface - GUI	52
1981	Space shuttle	179
1981	Stealth aircraft	258
1982	Aspirin - effectiveness	204
1982	Diabetes First synthetic insulin	195
1982	Genetic modification of a plant cell	17
1983	Blind signature	43
1983	Internet	53
1984	Computer - handheld	50
1984	DNA - fingerprinting	41
1984	LCD projector	32
1984	MAGLEV - world's first MAGLEV Train	262
1985	Live Aid	149
1985	Wealth creation	25
1986	Heart lung & liver transplant	198
1986	Monoclonal antibodies	183
1987	DNA - first conviction based on DNA evidence	42
1988	Firewall	51
1988	Luggage - tilt and roll	109
1988	Nicotine patch	200
1989	HTTP URLs and HTML	52
1989	Selective laser sintering	232
1989	Zip file format	56

1990 ARM Chip design	20
1990 Web browser	55
1991 Ant robotics	214
1991 Viagra	190
1991 Wind-up radio	235
1991 WWW – World Wide Web	56
1992 SMS – text message	35
1991 Ant robotics	214
1994 Amazon	19
1994 Biometrics – iris recognition algorithms	48
1994 CMOS	48
1994 DNA computing	40
1994 Quantum cascade laser	230
1995 DNA Database	41
1996 Cloning – first successful cloning	192
1997 Car – world's fastest	272
1997 Hard drives – for personal computers	52
1998 Bus – hybrid drive	271
2001 Bridge – tilting	73
2001 iPod®	54
2004 Bicycle – li-ion powered electric bike	269
2004 Graphene	220
2006 Battery – structural	123
2006 Car – world's fastest diesel	272
2007 Nanowire battery	229
2008 Face recognition	42

Chronological index – by year

2008	Laser timing	138
2008	Stem cells - first practical use	183
2008	Steri-spray - sterilization of water	211
2009	Aircraft - the world's first autonomous aircraft	255
2014	Brain -'locator GPS' Discovery	206
2014	Vantablack® - world's blackest material	86
2020c	Car - world's fastest future car	273
2020c	SABRE HOTOL - hypersonic jet/rocket	258

Printed in Great Britain
by Amazon